"It sparked a deep appreciation, inspiring me to salute the Earth as a great creature, a mighty evolving, growing, feeling and being organism with a life of its own."
Helen Nearing
co-author, LIVING THE GOOD LIFE

"Important reading for everyone... Once I picked it up I didn't put it down until I read the last sentence. Its critical topic is interwoven with the author's first-person accounts.
"Mike uses down-to-earth language and humor to tell a vital story. As a professor, I'm tempted to assign this book to all I meet."
Clifford E. Knapp
Outdoor Education Faculty
Northern Illinois University

"Delightful, illuminating communication at its best—containing stories and guidance for experiencing Nature in the full thrill of her presence. Playful, scientific, but not academic. Take it and read!"
Dr. Thomas Berry
Founder and Director
Riverdale Center of Religious Research

"This is not a reading experience, it's a hologram. It is communicating through talking leaves, something of what whales communicate to each other... you are there."
Robin E. Lagemann
Indigenous People's Activist

"One cannot—and should not—attempt to escape the depth of concern, commitment and feeling of this book. It offers great food for thought, approaching a banquet."
John F. Disinger
Professor
School of Natural Resources
Ohio State University

HOW NATURE WORKS

Regenerating Kinship with Planet Earth

MICHAEL J. COHEN, Ed. D.

Published by
Stillpoint Publishing
with
World Peace University
and
Center of the University for Peace, United Nations
Portland, Oregon 97215

STILLPOINT PUBLISHING
Books that explore the expanding frontiers of human consciousness
For a free catalog or ordering information, write:
Stillpoint Publishing, Box 640, Walpole, NH 03608 U.S.A.
or call 1-800-847-4014 TOLL-FREE (Continental US, except NH)
1-603-756-3508 or 756-4225 (Foreign and NH)

First Trade Edition

HOW NATURE WORKS

Copyright©1988 Michael J. Cohen, Ed.D.

All rights reserved. No part of this book may be reproduced without written permission from the publisher, except by a reviewer who may quote brief passages or reproduce illustrations in a review; nor may any part of this book be reproduced, stored in a retrieval system, or transmitted in any form or by any means electronic, mechanical, photocopying, recording, or other, without written permission from the publisher.

This book is manufactured in the United States of America.
Text design by Rostislav Eismont Design, Richmond, NH.
Published by Stillpoint Publishing,
a division of Stillpoint International, Inc.,
Box 640, Meetinghouse Road, Walpole, NH 03608.

Published simultaneously in Canada by
Fitzhenry & Whiteside, Ltd., Toronto

Library of Congress Catalog Number 87-062668
Cohen, Michael J.
How Nature Works
ISBN 0-913299-45-6

9 8 7 6 5 4 3 2 1

TABLE OF CONTENTS

INTRODUCTION

PART ONE: TOUCH THE EARTH
Chapter 1. The Mediated Pulse — 17
Chapter 2. Learning From Nature — 33
Chapter 3. New Vistas Dawn — 49
Chapter 4. Experiential Facts or Fantastic Fiction? — 63
Chapter 5. The Science of Gaia — 77
Chapter 6. Voice of a Life System — 107
Chapter 7. The Affinity Bond — 119
Chapter 8. The Secret Life of Planet Earth — 139

PART TWO: THE CIVILIZATION OF NATURE
Chapter 9. The Tropicmakers — 159
Chapter 10. Nature Abandoned — 177
Chapter 11. Recharting the Course — 197
Chapter 12. The Golden Glow — 217
Chapter 13. Dancing in the Green — 231

Epilogue — 253

Appendix A. Information Concerning the National Audubon Society Expedition Institute — 255

Appendix B. Information Concerning the Conference on Regenerating Kinship with Planet Earth — 257

Appendix C. Information Concerning Ownership and Protection of Rain Forests — 259

Bibliography — 261

DEDICATION

For Danny Miller and Rufus

Written with thanks to those who first sponsored this book's subject.

Brian Bedell, *National Audubon Society* • Robert Binnewies, *National Audubon Society* • Larry Brown, *Maine State Senate* Gene Boyington, *Cobblesmith Publications* • Larry Buell, *Association for Experiential Education* • Michael Caduto, *New England Alliance for Environmental Education* • Marshal Case, *National Audubon Society* • Doug Chapman, *Institute for Expedition Education* • Karin DiGiacomo, *Boulder College* • John Disinger, *North American Association for Environmental Education* • Johnny Fisk, *National Audubon Society* • Mary Huegel, *Lesley College Graduate School* • John Hunting, *The Beldon Foundation* • Myrtle Jones, *Mobile Alabama Audubon Chapter (NAS)* • Charity Kruger, *Aullwood Audubon Center, Dayton, Ohio* • Richard Mason, *Hobart-William Smith College* • Richard Murless, *Wilderness Southeast* • Nancy Neilsen, *Institute for Expedition Education* David Orr, *Meadow Creek Project* • Trudy Phillips, *Pennsylvania Alliance for Environmental Education* • Richard Schneider, *World Peace University* • Jim Swan, *International College* • Wanda Terhaar, *New York State Outdoor Education Association* • John Tishman, *Tishman Construction Company* • Joyce Wolf, *Jayhawk Audubon Chapter, Kansas (NAS)* • Dick Wylie, *Lesley College Graduate School* •

ACKNOWLEDGEMENT

The author extends many thanks to those who evaluated and edited this book: Raina Imig, Debra Latham, Seth Benz, Lloyd Scott, John Winn, Selden West, O'Ryin Swanson, Michael Vogt, Anton Grosz and especially Serena Lockwood, who also processed the manuscript.

All photographs unless otherwise noted are by the author from color slides. Black and white conversions were made by Favorite Studios in Jacksonville, Florida.

Much appreciated scientific guidance was provided by David Laing and James Lovelock.

Constructing New Life Ways, *Boulder, Utah*

INTRODUCTION

The last six months of 1959 found me, saw and hammer in hand, converting an old Vermont farmhouse into a rustic ski and summer camp. The building stood on a 70-year-old stone foundation dug into a 10,000-year-old glacial knoll. I spent portions of those months groping in a dark farmhouse room constructing new interior walls and doors. It was dark there because Joe, a long-overdue local carpenter, never arrived to install the window. Since he had not seen the light with respect to his job, neither did I.

One evening, upon returning with supplies from the lumber yard, I saw that Joe had finished his construction. Before me stood the house with its prominent new window and Joe smiling proudly. But the window was neither horizontal nor vertical. It sat at an ungainly, mind-blowing angle to the roof and cellar, and made the formerly old-time but stately farm building look ramshackle.

"What's going on?" I calmly asked Joe. He, too, was flabbergasted when he saw his ridiculous masterpiece from outside. It looked simply awful.

Upstairs we went and soon found the problem. Joe had used a level and installed the window true to the level's readings. But the level also showed that the house and foundation were on an angle. Against the forest and field background, you would hardly notice their slant. Only if you placed a marble on the floor would its rolling indicate the building's bent. You could ignore the marble's message while inside the house, but outside, in contrast to Joe's beautifully level window, the house became the leaning eyesore of Vermont.

We never corrected the house's bias. Joe came back the following day and re-installed the window crooked by making it parallel to the warped angle of the roof and cellar line. The house looked stately once again.

I relate this story because this book is about the bias of the strife-laden structure in which we are born and raised. As we grow up in our excessively violent, environmentally unsound modern society home, our ingrained thoughts and feelings bear the warp of its foundation. And as we reap its distortions, we foster its problems.

Level-headedness discloses that some of modern society's foundations are faulty, such as how we deal with our hurtful stress, peace and environmental problems. But rather than correct our fundamentally disjointed contacts with Planet Earth's nature, we hide them under language and relationships as crooked as my farmhouse window. That is modern society's angle.

Lying beneath our band-aid solutions, painfully writhes our desire to be true to the world and ourselves. Daily we watch Earth boil in response to our deceitful cover-ups and rejection of its harmonious ways. Our home is a disaster area. However, a springtime of awareness has arrived and it's time to straighten out the house.

Level-headedness says we must listen to the Earth if we wish to gain its wisdom. Otherwise, to our cost, we level the Earth.

Learning Earth's language takes time. It took me almost thirty years of year-round camping in the natural world to articulate this book's special message. This book levels with you and Earth because my experiences are true to the wilderness that maintains global life.

Three decades of exposure to natural systems, while isolated from modern society's bias, taught me how Nature works.

While living in Nature, I learned the value of sensations and feelings as well as that of rational scientific training. But only when Nature affirmed my sentient self did the source of modern problems become apparent. And as I identified the effects of modern society's biases on me, I straightened out with respect to the natural world.

In natural outdoor settings, inner feelings empowered my observations of Nature. These feelings pushed my observations through the thick walls that formerly limited my perceptions. I discovered that although great evidence exists to the contrary, modern society believes that for survival, it must abandon and excessively exploit Nature. We have built our global and personal lives on this fundamental aberration.

Our separation from Nature pervades all aspects of modern life. It weakens us. Our battles with Nature ignite the wars between and within ourselves.

To rectify our separation from the natural world, I have designed a practical blueprint of how Nature works. This map doesn't divide people from the global life system. Uniquely, it depicts both as one, and allows its users to rechart the stormy Nature-abandoned course of their personal lives. That is what this book is all about.

Those of us who use my map will approach personal and global sanity. Like using a level, the map shows us what our biased upbringing prevents us from seeing. It allows us to "true up" our distorted lives by building compatible relationships with Nature. I believe you will find it useful and deserving of your trust.

This book also presents the scientific findings of world-reknowned scholars that combine to sustain my conclusions, formed by years of observations made while living in natural settings. A suggested bibliography is in the back of the book, and all reference numbers in the text refer to the materials appearing in the bibliography.

A study guide is also provided at the end of each chapter to help you discover those personal experiences that give the subjects presented a greater, more meaningful depth. This study guide is particularly important, for you best know something by experiencing it. Books may help you to know life intellectually, but their message is vicarious. For example, many people who have read books on how to build furniture can't do it because they've never seen, touched or used the necessary tools and materials.

As members of modern society, at home and in school, we learn to label. In time, excessive labeling replaces our experiences. We get to know the song of our lives by its words, but without its music.

This study guide temporarily lets you gently remove some of the labels you use so that you may consider what lies hidden beneath them. It also offers you new experiences from which to learn. If you want to increase this study guide's value to you, do it with friends and share your perceptions with each other.

If you're an educator or person with further interest in teaching materials and techniques for this subject, contact the National Audubon Society Expedition Institute. The Institute is designing a workbook that further explores holistic learning exercises. Also, we would appreciate learning of your experiences in this area.

Many universities, organizations and conferences have welcomed this book's blueprint. To date, I have given it as a keynote presentation, workshop or seminar for the institutions listed below, and I thank these professional groups for their assistance. Time permitting, I welcome invitations to do the same for other interested parties and outings.

Institutions where this material has been presented:
- Pennsylvania Alliance for Environmental Education
- Kansas State College Biology Faculty
- Jayhawk Audubon Society
- New England Alliance for Environmental Education
- New York State Outdoor Education Association
- Association for Experiential Education
- North American Association for Environmental Education
- First International Symposium for the Promotion of Unconventional Ideas in Science, Medicine and Sociology
- Northeast Regional Conference, National Audubon Society
- Lesley College Graduate School Seminars
- Youth Environmental Society, Princeton, New Jersey
- Hobart-William Smith College
- Lee County Environmental Education Center

- Walden School
- National Audubon Society Midwest Regional Conference on Education
- IS THE EARTH A LIVING ORGANISM? Symposium
- National Audubon Society Bi-Annual Convention
- World Future Society
- First North American Congress on Environmental Education
- New England Outing Club Association
- Environmental Studies Graduate Seminar, University of New Hampshire
- Wilderness Southeast
- Oatland Island Environmental Education Center, Savannah, Georgia
- Ogeechee Audubon and Sierra Club Chapters
- Common Ground Symposium: Ecology, Culture and Imagination
- Mobile Alabama Audubon Society
- Mobile Alabama School District Environmental Center
- Wilderness Quest
- Brown University Symposium on Peace and the Environment, Meadowcreek, Arkansas

Some of the following material is drawn from my previous books: **Prejudice Against Nature, Our Classroom is Wild America, As if Nature Mattered** and **Across the Running Tide.** I feel honored that after two decades a publisher still finds them worthy. On the other hand, sometimes I get the feeling he's saying, "Dr Cohen, the environmental situation has not improved—this time, do it right."

Hands-on Knowledge, *Sanibel Island, Florida*

PART ONE

TOUCH THE EARTH

Emphasizing Symbols and Images, *Martha's Vineyard, Massachusetts*

CHAPTER ONE
THE MEDIATED PULSE

During the early periods of nuclear bomb testing, the United States Atomic Energy Commission (AEC) studied how much exposure Americans had to the radioactive particles that their experiments released in the air. They found that from the moment of birth to the day they die, average Americans spend over 95% of their time indoors.

Like too many government surveys, the AEC study confirms what you and I already know. We are born and raised indoors, artificially closeted from the natural world. Our closeted environment usually consists of the house, school, car, bathroom and office.

Our habits form from contact with these indoor surroundings, not from Nature's wonders. The indoors conditions our thoughts and feelings. It programs us. Our separation from the pulse of snow, wind, rain, temperature, clouds and sunlight robs us of their life values. We know the natural world not by our ingrained experience with it but by a mediated lifetime of words and images describing it.

Three decades ago I was a 29-year-old biology major and teacher. Falsely, I considered myself knowledgeable about natural system functions. Like so many other folks, with respect to Nature I lived in a vast, unreal dream world of scientific thinking, terms, diagrams and myths. My distorted view of Nature was only surpassed by my lack of awareness that I carried a distorted view. Like many others, I presumed that ecology was the study of echoes.

But that is no longer the case. For the past three decades, I've lived in natural systems and learned the impossibility of the natural world through language alone. It is like trying to explain the color green; the natural world is an event, a pulsating energy that can't be completely conveyed by symbols or images.

In these chapters and in the photographs of Nature's classroom that accompany them, I share my outdoor experiences with you. They

Learning By Doing, *Lehi, Utah*

explain how Nature works and how it can improve your daily life. They map out for you what our blinded upbringing makes us overlook, yet which sits right before our eyes.

This book is like the story of the customer at a Russian restaurant who calls over the waiter.

"Waiter," he says, "taste the borsht."

"You don't like it?"

"Taste the borsht!"

"Look, I got fifty-seven other customers here. I'm gonna taste everybody's borsht? I won't get any work done!"

The customer stands up. "SIT DOWN and TASTE the BORSHT!"

So the waiter sits down, looks around the table and says, "Where's the spoon?"

The customer points his finger near his head and says, "AHHHHH!!"

Like the role of the borsht in the story, this book leads us to things we don't ordinarily see. And as shown by the chaotic state of the world with respect to Nature and peace, our upbringing is a load of borsht and we are the beet generation.

In the chapters that follow, I create a no-borsht map. It shows how we can blend with Nature's ways.

At birth, Nature endows us with sensitivities to a vast spectrum of its manifestations including electromagnetic waves, sight, sound, touch, smell, taste, consciousness, imagination, memory, rational thought, intuition, hunger, thirst, love and lots more. As children of Nature, we inherit the ability to know the natural world in multiple ways. To adults, unfortunately, these inborn abilities often seem magical or unreasonable because during our upbringing we've lost our capability to use them.

For reasons we will discuss later, modern society emphasizes our symbols, our intellect and our rationale. Unused and unexercised, our other senses disappear. You can experience firsthand how our past upbringing and history dramatically influence our present perceptions of Nature.

Fill three bowls with water (or borsht): one very hot, one very cold and one lukewarm. For a minute, place one of your hands in the very hot water, the other in the cold water. After a minute, plunge both your hands into the lukewarm water; your sensations convey that the lukewarm water feels hot to the hand coming from the cold water, and cold to the hand coming from the hot water.

This simple experiment demonstrates that we know the present only with respect to where we've been. However, feelings from both hands, along with awareness of the experiment's total process, tell a fuller story. That's why we need a blueprint of wholeness for discovering Nature. Without it, past experiences distort present knowledge.

Knowing Nature only by reading about it limits our consciousness to the dimension of words and images. Alone, words, symbols and images mediate our experiences. They provide incomplete knowledge and can remove us from life's fullness.

In learning about how Nature works, we need a full awareness of mediation's limits in order to overcome its shortcomings. For example: Study Figure 1-1 below and note the phenomena that occur. Trust the impressions you have while observing the figure. You can validate them because they actually happen to you, not because they may or may not happen to others. Note also that these same phenomena occur in our relationship with Nature and with the land.

Figure 1-1

1) Like the global life system, the whole figure is just a complex design. It's not of your making until you recognize something familiar in it.

2) In the total design of the figure, at any given moment you can see either a white vase or the black faces surrounding it. That's like looking at a forest and seeing either trees or logs.

3) Look at the figure as a vase and note that the faces then disappear. Reverse this. Look for the faces and note that the vase disappears. See the forest as logs and you no longer see living trees.

4) You cannot see both the vase and the faces at exactly the same moment. You can only be conscious of one symbol at a time. When you mentally label and think about faces, you see only faces. Think about logs; you will see only logs.

As long as we label the picture, it becomes the label. The picture changes only as the label changes. It is either faces or a vase, not both; trees or logs, not both.

5) Once we label the picture, we lose the original integrity of its design. It becomes merely our perceptions of it. This should teach us that focusing upon one part of a total image causes the remainder of the image to appear as background. Furthermore, if we repeat and reward this over time, it becomes a habitual perception. Often, as we learn a label or term, we learn only its meaning in our society. We often see habits, not realities. That's how a beautiful wildflower becomes a weed, or why we are disgusted by helpful insects.

6) Look again at Figure 1-1 and place importance on the color black. Notice that when you look for black, you sense the color and at the same time you can see the faces. Color is a sensation attached to the faces. By focusing on the sensation, we sense it, but we don't necessarily symbolize it. This allows us to perceive awareness of both color (sensation) and image simultaneously.

In this same way, we know or relate to Nature by how we symbolize and/or sense it. For example, we might symbolize that a snake is harmless, but our sensations can fearfully color this knowledge.

7) The inability to see beyond our labels is also documented in the sad story of Amanda Brown, a highly religious woman caught in her house as floodwaters began to rise. "No thank you," she told the policemen who drove out to her farm to take her to safety. "The Lord will save me."

Words Focus Life Experience, *Taylor Slough, Florida*

Soon the water had reached the second story and rescuers appeared in a rowboat. Again she refused, reaffirming her faith in the Lord. Later, with only the roof remaining above the raging stream, a rescue helicopter threw down a rope ladder, but she clung doggedly to the chimney and again refused to leave. "The Lord will save me," she yelled and refused to leave her perch. Suddenly, with a wrenching of old timbers, the house was torn off its foundation and she was thrown into the torrent and drowned.

Awaking in Heaven, she was highly incensed that the Lord had taken her life, and she decided to express her feelings to Him in no uncertain terms. "How could you let me die," she asked, "after all the prayers and praises I've offered in Your behalf?"

"I don't see what you're so upset about," answered the Lord. "After all, I did send a police car, a rowboat and a helicopter!"

Notice how the important label "Lord" hid the rescuer's efforts.

When we view Nature's design using only words and images, the same phenomena occur as in relating to Figure 1-1. Our values, labels and imagery separate Nature into imaginary parts. They disconnect and hide the total design of the whole. In this way, we enter the picture and distort its design by mediating it with labels.

Multidimensional learning reduces the chance of distorting reality. The more knowledge, sensation and feeling we experience, the better we know an entire entity or situation. For example, if two real people were facing each other and we could talk with them, sense their presence in many colors, touch them, watch their reactions and movements, feel their emotions, hear them and know them in 3-D, we would not see a vase between them. Nor would they disappear if we momentarily focused on the space between them. They would more fully convey themselves because we would more fully receive them. That's why knowing Nature by experiencing it produces a fuller, more accurate picture of it than knowing it only intellectually.

Our upbringing emphasizes naming things, making money and spending time indoors. These important background values strongly influence our view of Nature. The following studies show some perceptual aberrations that occur because of modern life's emphases.

Examine this figure which has its own shape and integrity. Labeling the figure "glasses" or "barbells" often attaches the qualities of these labels to the figure. Asking people to draw the labeled figure from memory produces this figure from many who labeled the figure "glasses," and this figure

from many who labeled it "barbells." It is the label that has distorted reality. If a real barbell or eyeglass-type figure was presented, these distortions would be unlikely.[25]

Consider this example. For a week, a university field trip was required to study bird life in a canyon. The students then learned that the following week they would study reptiles in the same canyon. They were surprised because they had seen no reptiles there, only birds. But during the following week they observed many canyon reptiles, for their attention was now directed down toward the ground instead of up at the birds in the trees and sky. Meanwhile, recreational hikers and poets in the canyon during this same period reported that they always saw reptiles and birds as well as most everything else, including themselves.

* * * * *

Consider this story. A father and son were in an automobile that skidded off the road and overturned. The father was killed instantly, the boy badly injured internally. A prominent surgeon was immediately summoned to operate on the boy. Approaching the hospital operating table, the surgeon exclaimed, "I can't operate on this boy. He is my son." The father is dead; how can the boy be the surgeon's son?

Give up? Although this story was puzzling in the 1970's, today's consciousness makes it more obvious, for the key to the puzzle is that the surgeon was a woman, the boy's mother. Note, however, that if this story were a real-life experience in which you participated, there would be no puzzle to solve. Your senses and society would combine, and you would observe the absolute reality of the situation.

* * * * *

Why do most American rivers have power dams on them? It is because we can see wild rivers and also value them economically as power reservoirs.

This is similar to a game in which this figure a.

was placed on cards in a deck. When the players randomly selected these cards from a deck, they received money for each card. The players might also randomly select cards with this figure b.

Selecting this card made a player forfeit money. After the game was played for an extended period, the players were shown this figure.

Most of them easily recognized figure a., the money-winning figure, but figure b. was often very difficult for them to perceive.[25]

Our cultural values, you see, shape our perceptions of reality.

* * * * *

Here is an interesting and revealing exercise. Think of four-letter swear words you have heard. Observe that they usually describe acts of Nature. We usually associate negative feelings with Nature. Think about the natural life system within yourself, the biology that you inherited from the Planet: digestion, sex, wrinkles, body hair, feelings, sweat, excretion, etc. Your internal life system also is Nature and it, too, is viewed with a negative bias.

Another experiment demonstrates why we might feel more comfortable living in a town than in a forest. Meaningless Chinese-type figures of calligraphy were shown to a group of subjects. Some figures were shown only once, others as many as twenty-five times. The subjects were then asked which figures they liked. As expected, they consistently chose the figures to which they'd had the most exposure. The figures to which they'd had least exposure were least appreciated. This is because recognition of the repeatedly exposed figures had been rewarded in a subtle subconscious way.

You see, life itself is rewarding; it feels good to be alive. The unconscious reward, the joy of living, accompanied each interaction and was reinforced with repetition. Thus the very fact that we spend so much time living indoors tends to reinforce our tendency to want to live indoors. Rewards and repetition bias our outlooks. This might not occur if the figures used in the above experiment were real entities and the greatest exposure was to a cesspool. Sensing it would communicate other dimensions.

* * * * *

Once I lived part of a summer on an island. I didn't know the names of its plants, animals or rocks. They were only what they appeared to be as I sensed them. Then a naturalist visited and taught me their common and scientific names and their uses. Immediately I felt the difference. When the natural area became names, it lost some of its enchantment. Conversely, when I return to town after an extended wilderness trek, I find myself extremely taken by the written messages on billboards and in advertisements. It takes a while before I once again become numb to them. The abstract world of symbols and images is often a separate reality.

After thirty wilderness years, my consciousness of Nature is still distorted. Even when I objectively, unbiasedly attempt to select an undisturbed wilderness mountain as an example of Nature, I abstract it by placing many global processes, life-forces, functions and sensations in the background. And when I do, I lose sight of a great deal of what Nature is about.

For example, when I view the mountain for its economic values of lumber, minerals or recreation, I hide from view its importance as a vital part of the global ecosystem. When I label it scientifically, I often isolate it from the environment that sustains it. I also separate it from others when I use its esoteric scientific names. When I name the mountain after a hero of our society, I influence my perception of its natural integrity.

The greater familiarity I have with the mountain as a recreational, cultural or scientific resource, the more I appreciate it as such. The less I know of the mountain as a personal life relationship, the more comfortable I am with it as a negotiable label. Labels, however, often subdivide reality's wholeness and design.

My years out-of-doors show me that most of our symbol and image quirks are actually signals from our Inner Nature. The conflicts they contain are attempts by our Inner Nature to make us look further, to draw us into life's wholeness. Perceptual distortions urge us to discover how Nature works. They are trailheads to global coexistence. I have had the good fortune to follow some of these trails to their

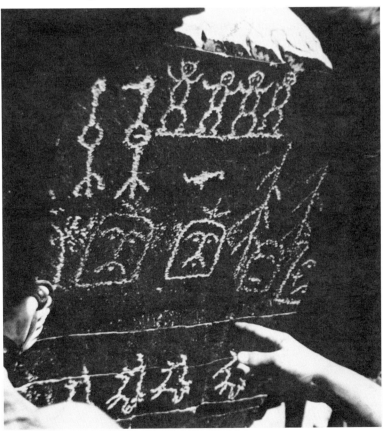

Discover Nature-Congruent Symbols, *Hopi Salt Trail, Arizona*

source, the vibrant global community that generates and sustains life as we know it. That is what this book shares with you.

For example, all of our grandiose conceptions of people's starring role on the Planet may be somewhat altered when we also view ourselves as a food source, grown and recycled by bacteria for their health and welfare. That's one side of the picture we seldom see.

How often reality fails to live up to its labels. Yet, as shown by the examples in this chapter, only when we are confronted with mediated life's distortions can we react to them.

The following chapters share hidden unmediated aspects of Nature that I've gained from label-confronting decades of tasty, colorful, cross-cultural, multidimensional sensations and experiences while exploring hiking, cycling, skiing and living outside. You, too, can find them through a form of learning that surpasses the limitations of the world of symbols and images.

This book augments and enriches our limited symbol and image notions of how Nature works. From them, we will map a more realistic view of our relationship with the natural world, a map my students and I use to incorporate Nature in resolving problems. With your permission it will serve you well, for as Confucius said, "The beginning of wisdom is calling things by their right name."

DRAWING CONCLUSIONS

Note: Each chapter that follows has a Drawing Conclusions section. You may best comprehend this portion of the chapters by coloring in the diagrams as indicated. Use red, blue, yellow and green.

This chapter indicates that because words and images are one-dimensional, they easily distort reality's design or cause portions of reality to disappear. We portray this by using the color red. Red catches our attention and warns of danger; and, of course, words are read, too. Color this diagram red.

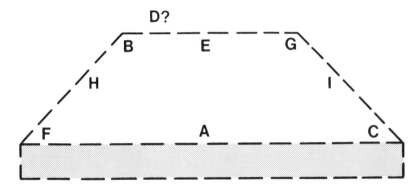

Figure 1-2

Figure 1-2 is modern society's life-guiding map of consciousness. Where is point D on it? The lettering sequence A,B,C,E,F,G,H,I suggests that "D" should appear, yet it can't be found. It could be anywhere. Using this map, we could be standing on point D or destroying it without realizing it. Actually, point D is the point where we live most harmoniously with Nature and with each other. It is the place where we bring to consciousness vital factors that modern society's symbols and images overlook.

STUDY GUIDE—CHAPTER 1
Most of this book's information was gained from living outdoors.

1) At what periods of your life have you been outdoors? What joys of life values did you find there? Have you retained them? Were they labeled? How?

2) What valuable natural aspects of your childhood have you lost as you've "matured"? ( )curiosity ( )desire for freedom ( )concern for living things ( )imagination ( )creativity ( )others (list them)

Your "lost" attributes still exist within you. How do you express them now?

3) Find and write out examples from your life for each of the perceptual phenomena that appear in this chapter. Identify some major labels in your life. What fuller experiences would you design to offset the bias you may have gained from labeling?

4) Try this experiment. With crayons, write a column of the names of ten colors in colors different than their names. For example, write the word "blue" with a yellow crayon, "red" with a green crayon. Then quickly look at the words you've written and say aloud the colors in which they are written. It's difficult because your Nature senses the color, but your acculturated symbolizing reads the words. There are members of many other cultures who would not have this difficulty, especially in cultures that have no written language.

5) Select any object around you and look at it. Then state what you see. Now touch your nose with your finger and look at the object again. Note that you see your finger and nose as well as the object. You could see your nose the first time you observed the object too, but it was "invisible" because this paragraph's first sentence focused your attention elsewhere. You and Nature are part of your life every moment of the day and night. Find other examples

of how you overlook your Nature by society focusing you elsewhere (how you feel is part of your Nature).

6) Traditionally, most education consists of learning words and symbols for that which we already know. Find five examples of this in your own education.

7) Lift the following common labels and see what you discover underneath them. Try to describe these entities in writing without using their names or the names of their major components: 1) tree; 2) cloud; 3) rain; 4) rock; 5) desert; 6) river.

NOTES

On this page or elsewhere, write down any thoughts or feelings that have come to you from reading this chapter.

Interesting Subjects Command Attention, *Silverton, Colorado*

CHAPTER TWO
LEARNING FROM NATURE

Sometimes the best way to deal with humanity's impact upon the natural world is to laugh about it. Humor helps to soften the pain of our destructive environmental and social situations.

The smog is so bad that this morning I thought I saw a blue jay. It turned out to be a cardinal holding its breath. And just last week I heard a Los Angeles mother yelling, "Junior! Don't stand outside. You'll get your lungs all dirty!" But it's wonderful how the scientists always think of new ways to protect us. Now there's a mouthwash for people who drink tap water.

But modern society is no joke. It's a way of life that the world has never seen before, and along with its special advantages, it presents plaguing conflicts. And no wonder. As modern Americans, we live in a society whose mainstream asks us to put ourselves on the line as educated citizens, skilled workers, economically sound adults, obedient children and responsible parents. As we do our duty to God and country, we strive for affluence, economic and social security, supportive relationships and a sense of place and community. We overload ourselves with responsibilities and the stress of "making it," seldom knowing where it's leading us, who's in charge or how to reduce society's negative impact on our complicated materialistic lives.

The rat race of new technologies, mortgages, insurance, family responsibilities and credit card payments crowds out an overview of our individual and collective lives. Nevertheless, most of us meet this enervating challenge with full vigor, the excessively "stressed out" state of the nation and Planet being the result of our dedicated efforts.

I have asked people at my workshops to identify the problems they face. Their responses give us an insight into some of the effects of our modern world.

"I think there's a lot more violence now than when I was young."
"The threat of nuclear war is most alarming."
"Runaway technology is ruining the land."
"I feel out of touch with Nature."
"My relationships are mostly competitive."
"Species extinction is rampant."
"We have alarming heart disease, drug and cancer rates."
"My boss makes me feel very uncomfortable."
"As a woman, I feel inferior and unequal to men."
"We have toxins in our food and soil."
"Families are breaking up. Divorce is practically the norm."
"We're still raping the virgin wilderness in the Redwoods."
"I feel bad about them making a shopping mall at Backbay Marsh, but it's perfectly legal."

Some people say that God creates a world full of problems in order to test us before entering Heaven. Whether you believe this or not, many other people, myself included, believe that in order to reduce our problems, it is we, not God, who must responsibly deal with them.

Modern society's dilemma, unfortunately, is that most of the deep joys we have either underlie or provide relief from the overwhelming problems we face. Is life simply coping with problems? If so, what hath God wrought?

* * * * *

As the previous chapter describes, in a mediated setting there are always two sides to everything. But, unless you spend much energy consciously selecting to psychologically and habitually live on the side that feels good to you, you get hurt. In my case it took thirty years to affirm that I enjoyed life more in living closer to the land than living in the city.

At the time of that decision, I was a biology teacher and guidance counselor who had accepted the position of Director of the American Youth Hostels in New York City. Living there, away from Nature, sometimes made me nervous and I found that city life often enervated and frightened me. If I let my guard down and felt attached to something, usually I received frustration from some disruptive or opposing factor. I was a success, but each day was often an unfulfilling series of compromises.

Impacts of Modern Life, *Rhyolyte, California*

For me, the straw that broke the camel's back was the day the President of the American Youth Hostels (AYH) blew his stack. As AYH's Executive Director, I'd previously convinced him and the AYH Board that the organization could make some desperately needed money by renting a large building in the mountains and converting it into a ski hostel. With much conflict and reservation, AYH plunked down rent and conversion money. Wouldn't you know, it didn't snow upstate that winter! We lost over $3,000.

But the President's aggravation with me was not limited to the organization's lost investment. No sir! This fellow was an accountant, a CPA, and I don't mean Compassionate, Pleasing and Alternative. He had yet another axe to grind. A discrepancy had been discovered in the AYH bank account. For some still unexplained reason, we had a surplus of over $6,000 in the bank. Nobody, including the bank, had any idea where the money came from. The only thing for sure was that it was ours. Somewhere, somebody had made a deposit slip.

The discovery elated me. Wonderful! The money more than covered the loss from the ski hostel catastrophe. But President CPA was livid. The money couldn't be explained, the books couldn't be balanced, nor the Internal Revenue Service satisfied. I told him to fudge the books and he told me I was incompetent. Finally I told him what he could do with the books—and you can tell the rest.

The incident typified my life in mediated America. I'd bet on Mother Nature's whims and sometimes get stung. I'd end up with a profit in a not-for-profit organization and rather than be celebrated, I'd be incinerated. Such was existence in mainstream's mediated, stress-filled house that so many of us call home.

It's one thing to get on an amusement park roller coaster knowing that the ride won't last long. But it's quite another to continuously live your daily life with disruptions often out of your control. You're never sure how long you'll last. I had been playing the game because I didn't know better.

Yet I did know that the natural world's design is wiser and more peaceful than anything else I'd known. In the out-of-doors, I felt more together, worked better with others and sensed more joy than in my familiar artificial world. And so I stepped off modern society's upheavals and decided to live on the side of the angels. It seemed more sensible to ride Mother Nature's full-dimensioned global pulse than modern life's potentially destructive roller coaster.

On May 8, 1959 I left my familiar indoor world of science, teaching, counseling, traditional music and administration in order to establish a new program allowing me to practice these same skills outdoors. I borrowed almost one quarter million dollars (in today's money) and founded an outdoor travel and camping program, becoming a wildlife sanctuary owner and director in the bargain. At age 29 I was going for broke, so to speak.

I knew that over 90% of new businesses fail. I had watched friends go under in their new endeavors, noting their stress and grief. Against doctor's orders, I strained my physical abilities to their limits and beyond, not to mention what I did to my personal economics and interpersonal relationships. In fact, I still carry the scars today.

Nature Evokes Peace, *Olympic Peninsula, Washington*

But though the venture's risk bothered me in my dreams, in the daytime I felt confident about leaving the mainstream world that I knew only too well. Where did this confidence originate? I gained it as a child. You see, I was born left-handed.

Being left-handed may seem like a small thing, but believe me, nothing is actually small, especially being a "lefty" in the 1930's. In my family, it was okay. But I was required to write right-handed in school and that felt wrong. In a "small" but continually aggravating way, I felt like minority people feel when required to do something that goes against their upbringing or natural features.

I struggled with writing right-handed. It made no sense to me, especially since adults applauded my batting lefty. I could hit a home run lefty, but I struck out batting righty. So it seemed batty asking me to strike out with respect to my penmanship.

However, it's one thing to read about my lefty dilemma and quite another to experience it, especially as a child. You can get close to how I felt by placing your tongue against the back of your upper front teeth and immobilizing it there. Now try to talk. Go talk to friends or strangers that way. It feels and sounds strange, doesn't it? And notice how they react. Now just as you feel you could talk O.K. if you used your tongue, I felt I could write O.K. if I used my left hand. But it wasn't permitted and so I spent five outrageous years learning to write "tongue-tied."

We are each born with "left-handed" qualities: literary, artistic, musical or some other special sensitivity that society doesn't necessarily nurture. For this reason, many of us never become the person we either could have been or wanted to be. My lefty experiences typify those of too many young Americans today.

I never wrote well. Constantly, "childishly," I argued my case for writing lefty because it felt "right." Finally, by sixth grade, they let me do it. By that time they'd let me do anything to move me on to the next grade. My writing was so bad, I could have become a doctor.

During sixth grade as a lefty, I surpassed the previous five years of right-handed penmanship skills. But I had not only learned to write, I had also learned that how I FELT naturally was right!

That lesson gave me some confidence. It taught me to honor my natural, lefty-type feelings, to understand and trust what they said. Since then, living in natural settings has further nurtured that seed of wisdom.

We only make something small in comparison to where our immediate values and directions place importance. Worst of all, overlooking the small things causes big problems. Can you relate my "lefty" situation to some element in your life?

Nothing likes to be considered small by the big picture. It's often very unfair because the big picture is usually nothing but small things that, for better or worse, rightfully or wrongfully, took over. Too many large-scale problems that the world faces today are nothing more than a number of small conditions all rolled up together.

Other circumstances related to school also validated my feelings for Nature. In third grade my class was subjected to a new experimental method of learning mathematics. Wouldn't you know that at the end of the year they found out that the experiment didn't work! Everyone in my class would be held back unless we learned regular math during the summer.

Generously, the school provides my class with a workbook and summer school tutoring for learning math. But my parents had already rented a lakeside cottage upstate. They offered to accommodate the school's tutoring requirement by assuming the responsibilities themselves. Given the chance, I doubt they'd do it again. Their vacation started when the summer ended.

All summer long, for two hours every morning, my parents and I plowed through the math workbook. Only after this exercise was I free to enjoy the lake and the countryside's delights. I was outraged into tantrums, and that summer left me with indelible feelings about schoolwork skills versus outdoor experiences. The former could be rigid, upsetting and tedious, the latter a joy. The difference wasn't academic, it was spectacular.

There was also the matter of the robin in the rat trap. A local merchant had set the trap and a robin's wing had gotten caught in it.

I walked by, was appalled and freed the robin. But feathers and bones lying nearby told a brutal story. So, every time I came by thereafter, I unset the trap, preventing it from injuring birds. One day I showed my mother the trap and told her of my activities. Unfortunately, the merchant overheard and he was furious.

Mother did not punish me as I expected. Instead she asked if I understood why the merchant was upset. I acknowledged his viewpoint, although I did accuse him of being stupid for hurting birds. After all, he could have just as easily put the trap inside where it might catch only rats.

The subject was dropped, and now that I think about it, I never saw that trap out there again. Perhaps Mom called the merchant and conveyed more politely my feelings on the matter. She was an activist before they coined the word, with a black belt in speaking. This and similar incidents further substantiated my feelings for wildlife. I felt terrible about killing an animal unnecessarily.

In retrospect, my early years could have been a life-experience course entitled "Natural Feelings Are Trustworthy Facts." That's why I could start my own program in 1959. I trusted how I felt about Nature and about being outdoors.

Since then, as part of the small accredited educational travel community that has been established, I've constantly camped out year round, sleeping in tents, sometimes on the ground, often under brilliant winter stars. As a critically thinking independent inquirer, I choose for my laboratory the face of North America, for my teacher the life experiences in the wilderness and sub-cultural settings which my students and I visit, from Newfoundland to California.

Unlike my traditional education, the natural world further encourages me to be rational, to respect all my sensations, thoughts, feelings and actions, not just those that fit into some preconceived mold. And though some people say I've been in the woods too long, the state of the environment tells me that too many of us have not been there long enough. From Nature, I have learned that most of our psychological, social and international unrest grows out of deep

Natural Feelings Are Valid, *Big Bend, Texas*

conflicts of our life habits with the natural world. We live as if the words "humus" and "human" don't have the same roots, which incidently they do. We are bewildered because, as the word means, we have "strayed from the wilderness."

Over a period of almost thirty challenging years, my independent outdoor program grew and became a success story as experiences like those described and pictured in this book became the B.S. and M.S. degree programs of the National Audubon Society Expedition Institute and Lesley College Graduate School.

As the photographs depict, on our expeditions and real-life encounters, as we camp out year round, we learn to emulate both the global life system's ways and the traditional knowledge disclosed by our backcountry visits. Our community survives by being open and honest and is much closer and more supportive than most families. For a year at a time, we learn experientially from our actions, senses and feelings. We learn from the people we meet, the places to which we travel, books in libraries all across the country, our classmates' and guides' reactions to us, our knowledge from previous years and the sensations of Nature.

We get to know each other better than we've ever known anyone, even our family. We discover new depths and meanings to ourselves. We learn peaceful coexistence by building it.

Far from a collection of unrelated academic subjects, our small consensus-based expedition communities organize their intense encounters to cover most disciplines. They find new values in living, learning and community, and although it's challenging, hard-but-fun work, they systematically explore the immense land and diverse people that make up America. (See the Appendix for further information about the Institute.)

From our travel and camping encounters, we learn that at birth our biological inheritance gives us many unnoticed mechanisms and sensations by which to know Nature. In 1965 a strange and wonderful event first brought this to my attention. I stood on a ledge overlooking the lower falls in Yellowstone Canyon, Wyoming. Directly below me, the Yellowstone River thundered over a high embankment and cascaded into the valley far below. My vision followed the foaming spray, moving with it on its showery journey into the valley. My eyes flowed with the river, almost as if the river directed them to do so. This is the way we watch most natural objects; the movement of our eyes automatically follows the flowing designs of trees, rocks, brooks and the flight of birds.

After a while, just for fun, I directed my attention to one particular spot in the great waterfall and just stared at that point for over a minute. During that time, the water flowed through my point of vision and I purposely didn't follow its flow. Then I lifted my eyes and looked at the rock wall alongside the waterfall. A startling apparition appeared; as if I were hallucinating, a section of the wall began flowing in a direction opposite to the flow of the river.

I repeated the procedure again, staring at a single point in the waterfall, letting the water pass through my point of vision for a minute or more, and then staring at the nearby rock wall. Once again the rocks flowed. I had my students try the same thing and the rocks flowed for them as well. They will do so for you, too, if you follow this procedure when visiting a waterfall.

The reason the rocks flow is because your eyes respond to your command to view a single point in the waterfall. Instead of naturally following the water's flow, some parts of your sight process continually pull back from their inborn, uniting flow with the water during the minute of staring at the single point in the waterfall. Over time, this pulling back becomes a conditioned act and your sight continues pulling back even when you look at a solid rock wall. This makes the rocks appear to move and distorts your perception of "reality." Try it and be amazed.

This "psychedelic" phenomenon is similar to the faces/vase phenomenon you saw in Chapter 1. In this example, once we place importance on some specific aspect of Nature, rather than on Nature's whole design and flow, our perception of the natural world becomes distorted.

I also find that as in following the waterfall's flow, the same phenomenon, in reverse, occurs when you join Nature's flow and sleep on the ground outdoors. There, your bedroom vibrates with Nature's pulse as the wind, wave and brook beat their song throughout the night. Trees creak, leaves rustle, twigs snap. While Earth bathes you in starlight, you resonate with the sound of the cricket and the mysterious hoot of the unseen owl. Flights of geese honk across the

Building New Roads for Education, *Dark Canyon, Utah*

dark sky, making the stars blink, and although your conscious mind may sleep through this enveloping music, your unconscious mind dances to it.

Strengthened over time, Nature's rhythm more strongly flows within you and you become more aware of it, its more vibrant existence allowing you greater consciousness of disruptions, such as how and when our society stops Nature's flow. Just as the strangely moving rocks stand out like a sore thumb, so do modern life's disorders, and you avoid them. And each evening you delight in returning to your exciting night dance.

All of us are constantly learning unforgettable, trustworthy facts and feelings from life. Sometimes they're unusual, but we should never ignore them. They contain thought, sensation and action realities that promote new awareness of both the natural world and our own personal growth.

Today, when I watch television programs about the areas I've lived in and loved for three decades, I don't recognize them. Watching those pictures on the tube is like saying you know what ice cream is because you've read a definition of it. Nonsense. Did you ever eat a definition? Was it chocolate or vanilla?

These days, normal relaxation methods don't fully work for me. While many people can relax by thinking about peaceful natural areas, when I think about them, I soon become stimulated by all the energies and excitement I've felt while visiting them. To relax, I count numbers backwards while visualizing them.

Scientifically, life experiences far surpass simply abstracting the way the world works, for their fullness conveys more truth about Nature than do mere words. They reinforce the value of life and, as the following chapters show, they open important vistas that might otherwise remain unrecognized. I guarantee that the methods, facts and findings you will discover within this book are more accurate, objective, testable, repeatable and complete than most of those that ordinarily guide your daily life. That has been my experience, and hopefully, through this book, you will allow yourself to make it your experience as well.

DRAWING CONCLUSIONS

Chapter 2 indicates some of the limits placed on our childhood consciousness. Usually modern society does not give much importance to the natural world, or to close community and feelings, in comparison to being indoors and using the 3 R's symbols and images. To bring this fact to our attention, we symbolize it on the map as point D. Point D gives us missing information from Nature, sensations and feelings. We may gain this information outdoors experientially.

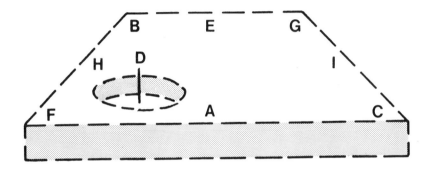

Figure 2-1

STUDY GUIDE—CHAPTER 2

The author finds that laughter helps him release stress.

1) Does laughter help you relieve problems? When? What kinds of things do you find funny? Can you locate the tension in problems that laughter releases?

People list the problems that concern them.

2) Which of the listed problems in this chapter bother you? What problems do you think are missing from the list?

3) Do the authorities or bosses in your life annoy you? What aspects of yourself do they agitate?

The writer selects to take the risks of living outdoors.

4) If you had five million dollars, how would you live your life differently? Why? What does the money buy?

Smallness is considered to be a comparative judgment.

5) What things do you consider "small"? What establishes their proportion? How could they become "big"? Who considers you "small" and how? What aspects of Nature are thought to be "small"? What is "big" about them? About yourself?

6) Which natural aspects of yourself do you feel have been stifled or embarrass you? In what ways do you consider yourself to be childish? If you lived in Nature, would your childishness be childish there? In what ways do you think natural systems are stifled?

7) Are there aspects of Nature of which you have been deprived?

8) In what ways do you trust your feelings? Where do you think your feelings should not be trusted?

9) Do you have a desire or sense of community? Is it being fulfilled satisfactorily? How might it be improved? Compare "community" in the natural world with the community you know.

NOTES
On this page or elsewhere, write down any thoughts or feelings that have come to you from reading this and previous chapters.

Experiencing the Best Teacher, *Colorado River Basin*

CHAPTER THREE
NEW VISTAS DAWN

A spectacle of cliffs and color mark the place where Bright Angel Creek joins the Colorado River deep in the bowels of Grand Canyon National Park, five thousand feet below the rim. The canyon's multibillion-year-old granite-veined schists portray the birth and changes of ancient landscapes and life. Here, on a scorching day in 1967, a group of kids, some critical thinking, a thunderstorm and a bag of potato chips cracked my thick walls of consciousness about life and the land.

The summer expedition was on a three-day backpack into Grand Canyon's inner gorge. Early that morning the red star, Antares, had brightly shone on our frost-covered sleeping bags at the canyon's nine thousand foot rim, but by mid-afternoon the inner canyon temperature, a mile below, simmered near 120 degrees.

To keep cool I hiked without a shirt on; it lay stuffed under my backpack's waist belt, preventing soreness and chafing from the twenty-pound load. Occasionally, I munched a potato chip to maintain my body's salt level in the broiling desert. Touching my bare stomach as I walked the suspension bridge that crosses the muddy Colorado River, I was surprised to find that though I was hot, my stomach felt extremely cold. I touched my arm and it, too, felt memorably icy.

Now a scientific person like myself has learned to explain the coolness generated by sweating through molecular theory: "Due to the sun's intensity, both the desert and my body heat up. However, in order to maintain its life-sustaining temperature, my body sweats, transferring its heat to sweat molecules, making them more active. They become vibrant enough to evaporate and jump off my body into the air, carrying away with them excess body heat while leaving a residue of salt on my skin."

With the molecular theory's explanation firmly in mind, I continued my journey through the canyon, giving the matter no second thought. It was a "small" matter in the infinite scheme of things. It never occurred to me that I was engaged in a miraculous event which kept me temperate while the rocks burned. I never wondered how my body knew when to sweat, when to stop sweating or how this information grew, was communicated or enacted. I was cool to the touch and to the subject.

Munching potato chips, I walked on, passing an ancient Indian site alongside the trail. These were Anasazi Indian ruins dating back prior to 1200 A.D. Next to the trail lay evidence that these people chipped flint and agate points, wove baskets and coiled clay pottery. I remained focused on these artifacts and their meanings. It never entered my mind that the Anasazi also sweated to keep cool. That was just a small thing I took for granted—the daily natural functions that kept them alive in this blazing canyon bottomland.

While we rested at Bright Angel Creek's junction with the Colorado River, I noticed storm clouds moving in from the southwest. "Now you're going to see something amazing," I told the expedition members. "As that thunderstorm pours, the raindrops will evaporate in the atmosphere's heat before they reach us here on the canyon floor. We'll watch the rain fall upon us, but we won't get wet." I knew that this fun thing was going to happen. It had happened many times before when I camped in the Grand Canyon.

A few expedition folks were skeptical, but all were excited by the prospect of the upcoming "dry storm." So the dark clouds came. And didn't they pour down torrents that drenched us to the bone while we stood there agape! "Hey, look everybody, I'm really dry," taunted one very wet young girl as she and the rest of the smirking group members scurried for raincoats and cover.

The Bright Angel's Call, *Grand Canyon, Arizona*

"Oh sure, Mike, it never rains in the Grand Canyon." How often those words would assail me that summer and in the years to come. Somehow, every student I worked with heard about the incident, and each felt urged to remind me of the event whenever I came close to suggesting what might soon take place. If I announced that we would have eggs for breakfast, the first response was, "Sure, and it never rains in the Grand Canyon." I couldn't even read tomorrow's tide or sunrise table without hearing not one, but a chorus of voices, blurt, "Sure, Mike, and it never rains in the Grand Canyon." Even today when I hear politicians and industrialists state how more nukes will produce peace, or how new toxins won't get into my water or food, I say, "Right! And it never rains in the Grand Canyon."

But the sky had opened up and rain it did. Quickly, the canyon had cooled. And, like soap rinsed out of your hair, the red sands and clays of Grand Canyon sloshed over the steep, thousand-foot black

Blood From Every Ledge, *Mammoth Springs, Wyoming*

walls of the inner gorge. The river turned from murky brown to deep, bloody maroon. Spectacular blood-red water poured from every ledge towering above us, spilling like paint dripping over its container's edge.

The incredible scene somehow looked familiar and suddenly I recognized why. I had seen it before in diagrams, movies and through my biology laboratory microscope while peering at animal capillaries emptying into veins. The flowing landscape looked like a giant vessel of the circulatory system gathering and carrying blood. It was as though we stood inside a living organism's bloodstream. What a blast!

Let me interject here that I am not prone to hallucinations and I never use drugs. It was no illusion. Everywhere the Grand Canyon countryside moved; it really looked as if it were alive. And as I stared, the staggering erosion and blood-red river soon triggered a number of questions that rattled my intellect and left me glassy-eyed with curiosity about this unbelievable spectacle.

—All these sediments and salts run down into the sea. Why doesn't the sea become increasingly salty?
—From natural erosion the ocean doubles its salt content every ninety million years. Why, after its many billion years' existence, isn't the ocean saltier than the lifeless Great Salt Lake in Utah?
—Does the ocean have some kind of kidney?
—Does the Earth sweat away salt as does my body?

Foolish questions? Just small things? They didn't seem so. The canyon's radical temperature change also brought to mind my cold, perspiring skin during the hike, just moments before. I wondered:

—Can the Planet become too hot or too cold for life to exist?
—Wasn't the Earth cooling itself by evaporating moisture the same way my body had cooled itself?
—If the Planet reacts like a living organism, does it also regulate its oxygen and carbon dioxide content as would another life form?
—If the Planet reacts like a living organism, does that mean that it is a living organism?

After a while, the raindrops falling on my head acted like buckets of water thrown in my face and returned me to the scientific consciousness of my upbringing. Although I had enjoyed my daydreams, I felt concerned about whether my mind was going. Maybe the heat was getting to me and a salt pill was in order. An occasional, "Hey, Mike, it never rains in the Grand Canyon," did nothing to help my self-credibility. To mention my Living Earth thoughts to others was the last thing on my mind. After all, it did rain in the Grand Canyon!

Had I been living in some human-created environment, the matter would probably have ended, relegated to the definition of daydream. But I was in the wilderness where Mother Nature continued to smile at me, and over the years she blew life into that spark of storm-ignited Living Earth consciousness. The laughing wind fanned the spark. An ember formed whose flame cooked up still more heated questions and inner conflict. "Was the Earth actually self-regulating and alive?" became a burning issue. The fire grew, as fuel for thought was added by both old memories and new experiences. What began as an academic conflict grew in scope and importance. It became a key to rectifying our posture toward the natural world, ourselves and each other.

Nature confronted my programmed outlooks continuously. For example, consider a wild scene in early March, our expedition's sixth month together. We hike to the summit of Everett Peak in Big Bend, Texas, where powerful updrafts blast through the forty-yard gap between the peak's towering pinnacles. We throw pebbles, sticks and branches into the gap, and the violent wind flings them upward in lofty arcs high across the mountaintop.

Giddy from exhaustion, we scream with laughter. Obviously Nature's governing body has met and repealed the law of gravity. "This one's for you, Ike!" yells Chris, hurling a rock down the roaring wind. Ike is Sir Isaac Newton, "discoverer" of the law of gravity by which objects in space are pulled toward the earth, not away from it. The rock hurtles toward the sky.

Similarly, we give tribute to Francis Bacon, Descartes and other seventeenth-century mechanists who reduced the pulsating life functions of Nature to inert, predictable mathematical laws and equations. "They remind me of the topsy-turvy bird," laughs Jim as a good-sized log goes sailing overhead. "Scientists have never collected a study specimen of it because it flys upside down. Every time they shoot one, it drops up." More laughter.

By consensus, the group solos near our Taconic Mountain campsite. Solos encompass the natural world. By sharing our reactions to them with each other, we become better acquainted. We disperse into woods, meadows, gulches and other selected sites and remain there for an hour. Afterwards, the reactions are varied. "I was scared"; "I identified six plants"; "imagined I was a frog"; "preoccupied with my grant application"; "a religious experience"; "frustrated not taking notes or pictures"; "removed my clothes"; "wanted someone there to tell me about it"; "a beautiful place"; "annoyed by an airplane"; and "heard music in the brook."

That evening we evaluate the solo. John said he enjoyed it but was uneasy because, "It didn't feel like school. What facts did I learn?" Many of us help him recognize how many different ways we did learn and how traditional education often devalues experiential learning. "You know how to masticate, without knowing how to say it, spell it or even what it means," laughs Bob, making the point.

Days later, a cloudless but cold morning finds the twenty of us sloshing through brush and sphagnum. The gooshy trail leads into a natural stadium whose playing field is a bog, a glacially formed pond so densely covered with floating vegetation that, skirting the few open water patches, you stagger across it feeling like you are walking on a leaking waterbed. The stadium's grandstand of increasingly taller larch and spruce trees circling the bog was once bogland, and these trees are the bog's future vegetation after it fills with peat. Some of us sink to our waists through the thinner patches of the floating vegetative mat. For the love of peat, mired in questions, we solo and then share our reactions and knowledge, using M. T. Watts, **Reading the Landscape** as a reference.

The expedition's discussion of the bog visit aggravates an intergroup conflict. The "mechanists" accuse the "tree huggers" of anthropomorphism as though it is a dirty word. They feel annoyed by those attributing to the bog scientifically unproven qualities like consciousness, sensitivity and communication. Their mainstream thinking takes lightly to unrepeatable facets of life that they can't model with computers. "But face it," argues Andy, "the bog was built before computers were around and it's probably never the same; it fluctuates. Computers can't even accurately predict tomorrow's weather, let alone this ecosystem's behavior. What's wrong with my impression that a bog is scar tissue grown by Nature to heal a glacial abrasion? Why ignore Nancy's idea that on some level all bog community members consciously, cooperatively organize and reorganize for their individual and collective survival?"

We travel to the Maine coast where the American continent and Atlantic ocean have a life-proliferating collision. Rocky precipices plunge into the crystal Gulf of Maine, forming spectacular oceanic microcosms in tidal pools. Armed with books and firm commitments to our group-unifying megaconsciousness, we backpack for five days along this coastal wilderness. We climb pristine examples of rock types which speak of ancient geologic forces, moving continents, melting rocks, calderas, metamorphics, erosion and rising coastline. Here, too, the sea whispers of our ancient kinship, telling us that it still runs in our veins. Periwinkles sing praises of their evolving independence into land forms. Tidepools reveal neighboring organisms ten feet from each other that can never change places without dying. Lampshells speak of salinity that has supported life for over 400 million years—never too salty, never too watery. Surging explosions of surf blend into immeasurable continental motions, while dancing winds whisper through dwarf spruce. Brightly painted algae and lichens beckon from their respective homes. Unique samples of life's spectrum crowd our minds.

We eat shoreline mussels, clams, periwinkles, fish, seaweed and plants. We drink from rainwater puddles, streams and rock seeps. Uncaged from traditional classrooms, nurtured by the living seacoast, our solos, meetings and seminars produce thoughts like: "Maybe we can't fully define life with words because, like this coastline and ourselves, life is an act. Since we are living beings, let's identify life by recognizing our life sensations and processes where they exist."

Over the decades, as I have lived with "Mother" Nature, she has constantly asked me to recognize how she and I share life and function identically. I guarantee she'd so affect any critically thinking, independent person spending decades in natural areas. For, though some people think it's a small thing, life in the natural world is a process different than its mediated abstract in our minds. Now, if that's a small thing, then so are the differences causing hostilities between nations, and the differences between living on the Earth and on the moon.

This I have learned from the time spent living in the very lap of Mother Nature. Most of the following places that I've visited have served as my home for a minimum of five days, and I've spent months in some of them.

NATIONAL AND STATE PARKS AND MONUMENTS

Gros Morne NP—Newfoundland
Otter Pt. Prov. Park— Newfoundland
Kejimkujick NP—Nova Scotia
Fundy NP—New Brunswick
Kouchibouguac NP—New Brunswick
Cape Breton NP—Nova Scotia
Roosevelt-Campobello NP—New Brunswick
Quoddy Head SP—ME
Baxter SP—ME
Acadia NP—ME
Monadnock SP—NH
Ft. Dummer SP—VT
Taconic SP—NY
Ft. Niagara SP—NY
Harriman SP—NY
Palisades SP—NY
Franklin Roosevelt SP—NY
High Tor SP—NY
Cape Cod Nat'l Seashore—MA
Rocky Neck SP—CT
Macedonia Brook SP—CT
Fahnestock Memorial SP—NY
Hudson Highlands SP—NY
Presque Island SP—PA
Assateague Is. Nat'l Seashore—MD
Prince William Forest Park—VA
Calvert Cliffs SP—MD
Shenandoah NP—VA
Cape Hatteras Nat'l Seashore—NC
Great Smoky Mtn. NP— NC, TN
Crater Lake NP—OR
Cascade NP—OR
Olympic NP—WA
North Cascades NP—WA
Warren Dunes SP—MI
Yellowstone NP—WY
Grand Teton NP—WY
Fossil Butte NM—WY
Devil's Tower NM—WY
Badlands NP—SD
Wind Cave NP—SD
Jewel Cave NP—SD
Bear Butte SP—SD
Rocky Mtn. NP—CO
Dinosaur NP—CO
Mesa Verde NM—CO
Great Sand Dunes NM—CO
Colorado NM—CO
Black Canyon of the Gunnison NM—CO
Cumberland Gap NM—KY
Santee SP—SC
Edisto Beach SP—SC
Lake Kissimmee SP—FL
Wekiwa Springs SP—FL
Everglades NP—FL
John Pennekamp SP—FL
Blue Springs SP—FL
Biscayne NP—FL
Organ Pipe NM—AZ
Saguaro NM—AZ
Chiracahua NM—AZ
Petrified Forest NP—AZ
Montezuma Castle NM—AZ
Walnut Creek NM—AZ
Sunset Crater NM—AZ
Grand Canyon NP—AZ
Rainbow Bridge NM—UT
Monument Valley Navajo Park—AZ, UT
Lehman Caves NM—NV
Yosemite NP—CA
Death Valley NP—CA
Joshua Tree NM—CA

Anza-Borrego Desert SP—CA
Kings Canyon NP—CA
Sequoia NP—CA
Lassen Volcanic NP—CA
Devil's Post Pile NM—CA
Russian Gulch SP—CA
Van Damme SP—CA
Redwood NP—CA
Prairie Creek Redwoods SP—CA
Richardson Grove SP—CA
Banff NP—Alberta
Jasper NP—Alberta
Assiniboine Prov. Park—British Columbia
Mt. Robson Prov. Park—BC
Yoho NP—BC
Kootenay NP—BC
Zion NP—UT
Bryce NP—UT
Canyonlands NP—UT
Natural Bridges NM—UT
Anasazi SP—UT
Great Salt Lake SP—UT
White Sands NM—NM
Carlsbad Caverns NM—NM
Bandalier NM—NM
Big Bend NP—TX
Big Thicket NP—TX

NATIONAL AND STATE FORESTS AND REFUGES

Plaster Rock Renous Game Forest—NB
White Mountain NF—NH
Mt. Mansfield SF—VT
Catskill Mtn. Preserve—NY
Adirondak Forest Preserve—NY
Ward Pound Ridge Reserve—NY
Housatonic SF—CT
Barnegat Nat'l Wildlife Reserve—NJ
Chincoteague Wildlife Refuge—VA
George Washington NF—VA
Blue Ridge Parkway—VA, NC
Pisgah NF—NC
Canaveral Nat'l Seashore—FL
Ocala NF—FL
Big Cypress Nat'l Preserve—FL
Loxahatchee Nat'l Wildlife Refuge—FL
Great White Heron Wildlife Refuge—FL
J.N. Darling Wildlife Refuge—FL
Okefenokee Wildlife Refuge—GA
Inyo NF—CA
El Dorado NF—CA
Sierra NF—CA
Gallatin NF—ID
Uinta NF—UT
Uinta Primitive Area—UT
Lincoln NF—NM
Moosehorn Wildlife Refuge—ME
Green Mountain NF—VT
Calvin Coolidge SF—VT
Mt. Greylock State Reserve—MA
October Mtn. SF—MA
Bear Town SF—MA
Martha's Vineyard SF—MA
Harold Parker SF—MA
Miles Standish SF—MA
Nickerson SF—MA
Sagamore State Beach—MA
Mt. Lemmon NF—AZ
Coconino NF—AZ
Tonto NF—AZ
Kaibab NF—AZ
Prescott NF—AZ
Lake Mead Nat'l Recreation Area—AZ
Glen Canyon Nat'l Rec. Area—AZ
Humboldt NF—NV
Ruby Mtn. Scenic Area—NV
Willamette NF—OR
Bridger-Teton NF—WY
National Elk Refuge—WY
Bighorn NF—WY
San Juan NF—CO

HISTORICAL PLACES

L'anse-aux-meadow NHP—NFLD
Saratoga NHS—NY

Hopewell Village NHS—PA
Williamsburg—VA
Jamestown NHP—VA
Yorktown—VA
St. Augustine—FL
Fort Bowie—AZ
Wilhelm Reich Museum—ME

Sturbridge Village—MA
Lowell NHS—MA
Springfield Armory—MA
New Bedford Whaling Museum—MA
Plymouth Plantation—MA
Mary Leakey Archaeological Site—CA
Pipe Springs—UT

SUBCULTURAL AREAS

Red Bay—Labrador
Francois—Newfoundland
Goose Bay—Newfoundland
Codroy Valley—Newfoundland
Burgeo—Newfoundland
Miramichi—New Brunswick
MicMac Reservation—New Brunswick
Chaco Canyon NM—NM
Zuni Reservation—NM
Taos Pueblo—NM
Tesuque Pueblo—NM
San Ildefonso Pueblo—NM
Santa Clara Pueblo—NM
Acoma Pueblo—NM
Appalachian Mtns—NC

Sabbath Day Lake Shakers—ME
Passamaquoddy Reservation—ME
Amish-Mennonite Country—PA
Koreshan Unity—FL
Immokalee (migrant workers)—FL
Navajo Reservation—AZ, UT
Hopi Reservation—AZ
Papago Reservation—AZ
Yaki Reservation—AZ
Wupatki NM—AZ
Canyon de Chelly NM—AZ
Havasupai Indian Reservation—AZ
Zia Pueblo—NM
Cochiti Pueblo—NM

MISCELLANEOUS

Squam Lake Science Center—NH
Felix Neck Sanctuary—MA
Audubon Schuylkill Nature Ctr—PA
Four Holes Swamp—Audubon—SC
New Alchemy Institute—MA
Kitt Peak Observatory—AZ
Rural Education Ctr—NH

Scarborough Marsh—ME
Sharon Audubon Sanctuary—CT
Patuxent Research Ctr—MD
Corkscrew Swamp—Audubon—FL
McDonald Observatory—TX
Wilson Observatory—AZ

Only personally experiencing these places can fully introduce you to what you need to know about them. But I have designed this book to teach you what my time in these areas has taught me about Nature's workings.

In the chapters that follow, if you permit Nature to touch your thoughts and feelings affectionately, she will in time kindle your Living Earth imagination and creativity. You, too, may enjoy recognizing yourself as her and vice versa. It's only natural.

Self-Regulating Bodies, *Big Bend, Texas*

DRAWING CONCLUSIONS

As exemplified by the idea that Planet Earth is a Living Organism, thoughts, sensations and feelings drawn from Nature may be outside modern society's ordinary map of consciousness. We can depict this at point D by having it indicate that a blue map of Nature's reality sits below our modern map of consciousness. Point D is a hole in the red map. It lets us see through it so we may observe Nature's map which I color blue, the color of the sky and Planet. The blue lines indicate where Nature may be found, but not its consistency, shape or texture.

STUDY GUIDE—CHAPTER 3

An unusual rainstorm embarrasses the writer, but leads him to think that the Earth might be alive.

  1) *Why do you think we become embarrassed when our words or predictions don't take place?*

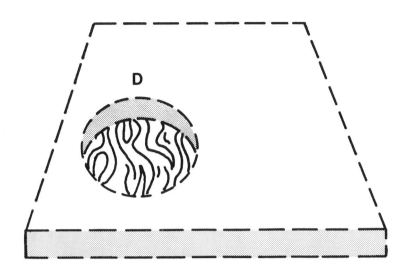

Figure 3-1

2) Have you ever thought or heard that Mother Earth was actually a living entity, not just a metaphor?

3) Did a specific person, book or situation teach you that Planet Earth was a non-living entity?

4) Do you entertain any wild thoughts that you believe are sheer fantasy? Do they have any possible basis in reality? Do you think other species have such thoughts or feelings?

5) What advantages do you think there are to modern society believing the Earth is a non-living entity?

NOTES

On this page or elsewhere, write down any thoughts or feelings that have come to you from reading this and previous chapters.

Biospheric Communication, *Mt. Mansfield, Vermont*

CHAPTER FOUR
EXPERIENTIAL FACTS OR FANTASTIC FICTION?

Contrary to popular belief, the most important question of our time is not the scientific accuracy of the information we receive. It is whether the state of the world and our lives satisfy us. Are we happy with modern destructiveness? I, for one, am not.

During the past three decades, by living in natural areas as part of our caring, "tough love" community, my students and I have been privileged to see our ingrained upbringing from a less artificial point of view than would have been the case otherwise. On our expeditions and elsewhere, this viewpoint has enabled us to live more harmoniously with each other as well as with the natural world. Nature has taught us to recognize that a lilac is blue against a red background and red against a blue one. Similarly, we have learned from each other that our modern upbringing is also *our* background, and that it acts accordingly.

Like a fish born in a brook, we are born into our mainstream view that the Earth is inert. Until our "red lilac outlooks" are confronted by other backgrounds, we remain set in our notions and their gnawing problems. We inherit culturally the scientific belief that life has no direction or purpose. That conceptual scheme pervades our institutions, and they, in turn, reinforce it. But, as for men not entering the women's bathroom, it's more of a habitual doctrine than a survival necessity.

By absorbing many viewpoints on our travels and engaging in multidimensional outlooks and processes, the National Audubon

Society Expedition Institute has uncovered a faulty, destructive, "dead planet" foundation of thinking that underlies mainstream's way of life. But by questioning it and dealing with it, we've learned to avoid many of mainstream's hurtful effects. And we've taught others to do the same.

The Earth acts like a living organism and is not, as we have been taught, a dead, inert, mechanical collection of minerals. That life and death difference sheds a new light on many relationships existing on the Planet as well as on the very nature of Nature itself.

Yet, though the concept of a living planet is scientifically controversial, every known peaceful, environmentally sound culture throughout history embraces a strong Living Earth belief system. I now understand why. Beings who depend upon each other for survival seldom attack each other. Their survival needs dictate that they cooperate or join together instead.

It is no accident that in the 1500's, the Living Earth conceptual scheme disappeared from our culture's consciousness, while at the same time, our modern problems started. We can cut, blast and plunder a dead planet with ease, but not a Living Earth-Mother that nurtures us. Instinctively, we respect that a mother knows best, and even humans don't normally bite the hand that feeds them.

Unlike a dry, classroom geology course, my Grand Canyon encounter had filled my head and startled me, leaving little choice but to get involved with the Planet's aliveness. For example, tasting the salt residue left on my skin from perspiration suddenly brought scientific thoughts to mind of the puzzling Colorado River sediments. "Why don't these abundant sediments make the seas too salty for life to exist in them?" A salient question, without doubt!

Then we'd visit geologic salt deposits and learn that they form from the evaporation of oceans: huge underground salt domes deposited by the evaporation of ancient inland seas—warm shallow inland seas formed from the rising and falling of the land— land whose upheavals were caused by the movement of the continents.

"Oh! So that's where the excess salt goes!" And the idea crosses the mind that perhaps God really did create these salt domes other than as a place for depositing radioactive wastes.

The Living Earth perspective urged me to ask the unthinkable. Over the aeons, did the continents move in order to produce evaporating seas, which maintained ocean salinity levels that would support life? Was the maintenance of salinity levels the reason that coral reefs formed warm, evaporating lagoons? Did the needs of sea life somehow change the globe's geography so that life in the sea could continue? Were moving continents and coral reefs' growth like the kidney of a Living Earth organism? Such fun questions never would have entered my "dead planet" mind before the Grand Canyon episode. Further living in exciting geological areas only encouraged these thoughts to grow.

My Living Earth episode also touched me when the weather became too hot or too cold. As I put on or removed clothing, I found myself having thoughts like, "How could the Planet regulate its temperature so that it might not become too hot or too cool for life's existence?" Then I gained awareness of atmospheric carbon dioxide and its heat-trapping greenhouse effect.

Carbon dioxide acts like a blanket. For temperature regulation, it can be added to or deleted from the Earth's atmosphere by the actions of plants and animals. Sea shells, for example, are made of lime and contain carbon dioxide as calcium carbonate. Did temperature changes affect sea shell-bearing microorganism populations? Did dying organisms store carbon on sea floors and underground as limestone, tree roots, peat, coal and oil to regulate the Planet's temperature for their own survival? Did they remove excess carbon dioxide when the Earth got too hot?

I enjoyed thinking this way about Nature even though these new ideas went against everything I learned at school. In academia, the Planet consisted of animal, vegetable and mineral. Animals and vegetables were alive, minerals dead. We were taught that life survives by adapting to the dead, static conditions of the Planet, not by organizing itself or controlling its environment. Scientifically, it was blasphemous to think that life had a purpose or direction.

But academic scientific thinking has been wrong many times before. People have thought the world was flat, continents didn't move, the Earth was the center of the universe and molecules were the smallest units of matter. They "proved" the moon's surface was covered with exploitable oil or thick dust. They "discovered" that at least half of Mars was covered with green algae. No, scientific thinking is not complete just by definition.

On the other hand, why should I pay attention to myself? I was no practicing research scientist. In fact, I came to live outdoors to get away from the academic world. Well, I'll tell you why I paid attention: just as being a "lefty" felt right, so did using the Living Earth concept. And persistently, Nature kept invading my mind, delivering the message that "Earth lives!"

Rock Waves Modify Memory, *Arches National Monument, Utah*

For years I never told anybody about my thoughts; I felt it was too risky. After all, credibility is survival. In my recent workshops, however, people told me that at one time or another they too believed the Earth might be alive. But like the Rabbi who played golf on the Sabbath and hit a hole-in-one, they couldn't tell anybody about it either.

Some people say I was "channelling" or had a "vision" in the Grand Canyon. That doesn't help. Where I come from, the scientific word for vision is meshugga, and that, in case you don't know, means crazy.

A more modern explanation arises from recent studies which find that rocks under compression emit long waves. These waves have been shown to influence electrically sensitive memory connections in the human mind. Thus, while exposed to compressed rocks, what were "rock-solid" memories, facts or personal experiences may become less certain. When you come right down to it, the scientific explanation doesn't really say much more than "vision" and, in the long run, is only a fancier way of saying "meshugga."

As I lived outdoors, Nature never let up on me. When feeling faint from lack of oxygen while hiking at 13,000 feet in the Rocky Mountains, for no apparent reason, I thought of an oxygen demonstration I had done as a classroom teacher.

It involved placing a glowing splint and a wet splint in a jar full of oxygen. The glowing splint would burst into flames that burned the wet splint. This showed that even wet wood could burn in the presence of higher oxygen levels.

Clearly, therefore, if the Planet's atmosphere contained too much oxygen, wet forests would burn and disappear. If too little oxygen were present, life couldn't survive. But the fossil record shows that neither event has ever occurred. Once created, forests and life always persisted. Like the development of an emotionally secure child, this was no accident. The oxygen level had to stay within the limits that supported life, otherwise life's record would have disappeared.

How might a Living Planet regulate its oxygen supply? Our visits to bogs as well as to coal and oil deposits suggested the answer. In these places plant carbon is buried. It doesn't enter the atmosphere.

Let me explain: Normally, plant photosynthesis separates carbon dioxide into carbon and free oxygen. The carbon becomes the plant's wood structure. Plants die and bacteriologically recombine with atmospheric oxygen, making carbon dioxide. But that heats the Planet through the "greenhouse effect."

What the Planet can do to prevent this is keep carbon and oxygen separated by physically keeping them apart. Thus, geologic processes bury some carbon as peat, coal, oil and limestone. Deep in the Earth, carbon cannot recombine with atmospheric oxygen. The Planet's procedure is like keeping Romeo away from Juliet.

In this Living Earth scenario, "dead" geologic rock formations and continental motions help organize the atmosphere in order to maintain life-supporting temperatures. To some people, that's also blasphemy. But is this relationship any different from that between "dead" bones, "dead" teeth and "living" lungs? Aren't they all part of our body's life design?

Other wild thoughts came to mind, such as: How might the Living Planet's systems intercommunicate what is needed when and where? Doesn't this hint of some internal consciousness? And how do people fit into the planetary scheme of life? Don't person-Planet life interdependencies exist? And if both people and the Planet are alive, can people learn about Nature by learning about their personal biology?

These thoughts evolved over a period of eleven years, during which time I never fully interpreted the Planet as a living organism. I would merely raise that possibility as something unthinkable to think about.

Within me, however, the intriguing possibility hit home. Not only my unusual exposure to Nature but also my contact with Native Americans, naturalists and radical ecologists further encouraged this heretic Living Earth thinking. Yet I never took the subject seriously enough to read about it. In fact, it never even entered my mind that anybody had ever written about it. It was never part of my education nor of anybody else's I knew. To me, it seemed more like a Peter Pan fantasy than reality.

But, unbeknown to me, during this same period, a British scientist, James Lovelock, came to the same strange conclusions about the Earth's life as I did, though his source of information vastly differed from mine. He had examined and questioned the "impossible" disequilibrium of gases in the Earth's atmosphere and concluded that only if the whole planet was live could its anti-entropic atmosphere exist. Our paths would cross, a decade later.

Meanwhile, because of its demonstrated educational and community values, my summer program had developed into a full year travel-study high school, college and graduate school alternative. By 1980, in fact, it was a B.S. and an M.S. degree program of Lesley College, as well as one of the educational programs offered by the National Audubon Society. I wrote accredited Living Earth courses and the book **Prejudice Against Nature: A Guidebook for the Liberation of Self and Planet.** Both the courses and the book reflected my encounters with the fact of the Planet being alive.

Self-Organized, Self-Preserving, Regenerating,
Schylkill Valley Nature Center, Pennsylvania

Our sponsoring institutions questioned these publications and the school's hands-on outdoor learning programs. I was saying that, to some extent, life itself was regulating all aspects of the ecosystem, such as acidity, regeneration and plant and animal distribution; all participated in organizing the global environment. I stated that plants live in certain areas because on some level they want to! It just feels right! "Are we giving degrees in Environmental Education or Fantastic Fiction?" asked our sponsors.

I responded to the credibility gap by conceiving an international symposium to address the controversy. With the dedicated assistance of Dr. Jim Swan and the Institute's staff and students, we called the question by producing a 6-day Conference at the University of Massachusetts in August 1985 entitled IS THE EARTH A LIVING ORGANISM?

It was like taking the lid off a pressure cooker. The scientific, interdisciplinary and cross-cultural response was nothing short of spectacular. It is conveyed in the lively Conference proceedings report that examines most aspects of the subject and its implications and is summarized in the next chapter.

Nature had taught me experientially about the Planet's aliveness by touching my thoughts, feelings and actions. But as the following *Environmental Information Magazine* review of the Conference shows, I was neither "touched" nor alone in my considerations of the Planet's aliveness. This time it didn't rain in the Grand Canyon.

> Gaia, the idea that the earth is a living organism, is an ancient belief. It is a concept that has been encoded in the ritual and culture of both native people and ancient civilizations, and has been put into practice in sacred architecture throughout the world. In the 1970s, British geochemist, James Lovelock, revived this ancient concept. He formulated a hypothesis to account for certain factors that provide long-term stabilizing forces on the planet, even after cataclysmic events. He suggested that the biosphere in its totality—soil, water, air, plants, animals and humans—can be described in terms of a cybernetic, self-regulating, living organism.

He gave the name Gaia to this hypothesis, after the ancient Greek goddess of the earth. His 1979 book, **Gaia: A New Look At Life On Earth**, brought forth current scientific evidence and theory supporting the idea that man is a microcosm within the larger living system on our planet. As science merges with the perennial wisdom of the earth, the acceptance of the paradigm of a living earth has profound social and cultural implications.

The full proceedings of papers presented at this 1985 Conference are the most complete collection of works on the Gaia hypothesis. It is a massive (4½ lbs.) collection of paper that describes the concepts of planet stewardship, holistic ecology and global consciousness. The papers provide modern facts, theories and functions that support the living earth concept and its striking implications. The work introduces new ideas corroborating the earth's biosphere, geology and cultures as an integrated, self-

Gaia: A New Look At Life, *Falls Island, Maine*

organized life system, displaying survival consciousness and homeostasis.

Its 1,324 pages include contributions by James Lovelock, George Wald, Mary Catherine Bateson, Charleen Spretnak, William Field, Lewis Thomas, Christopher Bird, Donald Michael, John Todd and 104 other credentialed scholars and researchers from the fields of physics, chemistry, anthropology, biology, philosophy and theology.

The Living Earth Conference did more than just help substantiate the idea that the Planet might be alive. It also verified the cooperative peace power inherent in that idea. Seldom do you get together a cross section of modern society's isolated, entrenched disciplines without strife, yet that occurred during the Conference's six-day duration. As if they were on our expedition, the presenters expressed their findings and experiences to the point of credibility risk and beyond, as far as words could take them. To their delight, they found that while they were "scientifically" defenseless, rather than being attacked by opposing forces or points of view, harmoniously, the opposition welcomed each presenter as a new area for consideration.

By recognizing themselves as interdependent participants of the Planet organism, the vastly differing Conference presenters established space to develop trust and cooperation where none ordinarily existed. Paul Winter sealed the bond musically. It should happen at the United Nations, too.

My only disappointment with the Conference was that it ended by labeling the Living Earth Organism, "Gaia." That esoteric word excludes billions of the world's people, including myself, who already have experienced its meaning. I've met many people who would attend Gaia conferences and could be excited about the Living Earth phenomena. If we don't include them, we cause the problems we are trying to solve. But sadly, language symbolism has once again taken power.

"Gaia," you see, labels the Earth organism as a motherlike female Greek goddess. The term excludes people who don't experience our

Planet religiously, who don't relate well to females, who are not academically aware of or enamored by Greek civilization, and who have trouble with "Mother Earth" because they've outgrown their mothers. Some presenters recognized this problem, but it appears that "Gaia" is here to stay.

Life is more than our mental or written word-image maps. It is the total experience of creating and sustaining local and global survival relationships through thoughts, feelings and actions. Life falters in their absence, like a gobble without a turkey.

Our upbringing's word-image map contains part but not all of the global life process. The increasingly chaotic state of the world testifies to the flaws in our conceptual schemes. To make them harmonious, we must incorporate the knowledge that the Earth functions like a living organism. That helps us to know the full design.

Unlike our present "dead-planet" belief system, a valid Living Earth conceptual scheme triggers additional feelings, actions and relationships. These help nullify the life-destructive aspects of our present consciousness and behavior. Had Moses known that some people would consider the Living Earth concept as blasphemy, there positively would have been another commandment.

DRAWING CONCLUSIONS

The blue map (which we can see through the red map hole at point D) is the Earth itself. Because the Earth is a living organism, our ordinary maps of the Planet will not suffice here. They do not indicate that the Planet is alive. This diagram indicates the Earth's position and the fact that its life supports us and our consciousness.

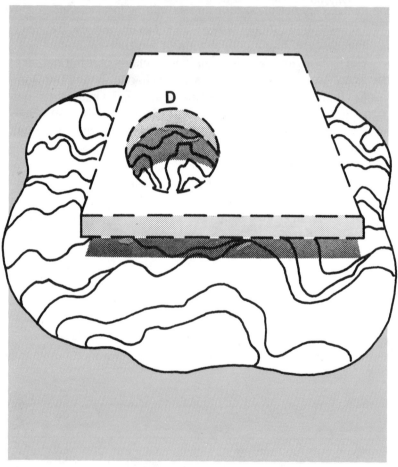

Figure 4–1

STUDY GUIDE—CHAPTER 4
Planet Earth shows symptoms of acting like it's alive.

1) What have you learned is the guiding force that permits life to exist on Earth?

2) What phenomena do you think regulate the Earth's temperature? salinity? atmospheric content? acidity?

3) Why do you think life as we know it exists on Earth, but not on our neighboring planets and satellites?

4) Have you ever thought that plants and animals might have feelings or communicate?

5) Why do you guess that most other cultures believe the Earth is alive?

6) What do you believe is the difference between life and death?

7) What might be different about our society if we knew Planet Earth to be a living organism?

NOTES
On this page or elsewhere, write down any thoughts or feelings that have come to you from reading this and previous chapters.

Wisdom of the Ages, *Flamingo, Florida*

CHAPTER FIVE
THE SCIENCE OF GAIA

Imagine the following scenario:

In his laboratory, a mineralogist shines his sun lamp on a microscope slide containing brown planet-like rock samples. Like an astronomer observing the planets through a telescope, he peers through his microscope at his wondrous rocks that behave like planets.

Scanning the slide in the sun lamp's light, he detects amongst the brown rocks an unexpected new object. It is a globular, rotating blue rock consisting of the minerals found on Planet Earth.

As do the brown rocks, the blue rock has an atmosphere as well as the motions of sub-atomic, atomic, molecular and chemical activities. But throughout his experiments, the blue rock alone maintains moisture-laden clouds, oceans and continents.

Unexplainably, the scientist finds free oxygen gas and chlorophyll on the Earth-like blue rock, but not on the brown ones. Yet, unlike its brown counterparts, the blue rock rejects the addition or subtraction of oxygen gas to its atmosphere.

He adds water to the slide and increases the sun lamp's intensity. The brown minerals, no matter what their distance from the sun lamp, increase in temperature and their water boils away. But, although it sits amid the brown rocks, the blue rock's color lightens, its gas ratios change, and its temperature and water level remain constant. The scientist adds salts to each rock's liquid areas; all increase their salinity except the blue one.

His energy measurements disclose that, unlike the brown minerals, the blue rock continually ingests high quantities of light energy and emits low energy quantities. But, upon turning off the sun lamp, the blue rock's motions cease, its free oxygen disappears, its atmospheric gas ratios change, large amounts of carbon dioxide appear and it turns brown.

The scientist's final report states that in the presence of sunlight, the blue rock self-regulates its temperature, atmospheric gas ratios, salinity and chemistry. It functions like a warm-blooded plant cell, like a living organism.

He observes that in the long run, all the entities of the blue rock, like organelles in a cell, interact as a whole. Only when he separates them or observes them individually for a short time, do they assume different properties including life and death. Long term, in congress, they appear to sense and commmunicate, thereby sustaining the optimum environment for their cell's life.

The scientist concludes that, like a virus, the blue rock is mineral, but it is also alive. It appears to be a hologram of the third planet from the sun. HE NAMES THE BLUE SUBSTANCE EARTH, BECAUSE IT DOES WHAT OUR PLANET DOES.

The above fictional dramatization encapsulates recent Gaia Hypothesis' scientific thinking. It is a composite of the findings and opinions of some of the world's finest scientific minds.

This chapter presents edited quotations from others' work, not my own. Based on ideas from the proceedings of the Audubon Conference IS THE EARTH A LIVING ORGANISM?[6], as well as on other publications listed in the bibliography, this chapter gives a sampling of the 1987 status of the Living Earth concept. It concurs with my experiences as described in this book and also conveys why, historically, the Living Earth concept disappeared from Western culture.

Historical Overview

We presently live in a state of disharmony with our Planet. It is a responsibility of science to investigate the possibility that our lives don't synchronize with the Earth because, although it is a living organism, we perceive it to be dead matter and hurtfully treat it as such.

Yet as far back as the written and spoken word allow us to go, some learned, and most "illiterate," people have considered Planet Earth to be Mother Earth, a living organism.

Black Elk: "Is not the sky a father, and the Earth a mother, and are not all living things with feet or roots their children?"

Rolling Thunder: "The Great Spirit is the life that is in all things—all creatures and plants and even rocks and the minerals. All things—and I mean all things—have their own will and their own way and their own purpose."

Lame Deer: "You have despoiled the Earth, the rocks, the minerals, all of which you call dead, but which are very much alive."

Aldo Leopold: "The land is an organism."

Lewis Thomas: "Viewed from the moon, the astonishing thing about the Earth is that it is alive . . . Beneath the moist, gleaming membrane of bright blue sky, it has the self-contained look of a live creature full of information, marvelously skilled in handling the sun."

John Lobell: "The Earth is a living organism capable of the self-regulation necessary to maintain the temperature range and complex chemical balances that support life."

Noel McInnes: "Earth is an organic spaceship. It lives."

Wendell Berry: "Our bodies are not distinct from the Earth, sun, moon, and other heavenly bodies."

Elizabeth Drew: "Sown in space, like one among a handful of seeds in a garden, the Earth exists."

Harriet C. Childs: "Oh beautiful Earth, alive, aglow."

Job 12:8: "Or speak to the Earth, and it shall teach thee."

M.C. Richards: "The world is alive throughout. Even the processes of death are connected in the ongoing movement of human being and universe."

Gary Zukav: "We would like to think that we are different from the stones because we are living and they are not, but there is no way we can prove our positions. We cannot establish clearly that we are different from inorganic substances. Inorganic substances can make decisions, react to stimuli and process information."

Guy Murchie: "From space, I get the definite, but indescribable feeling that this my maternal Planet is somehow actually breathing—faintly sighing in her sleep—ever so slowly winking and wimpling in the benign light of the sun, while her musclelike clouds writhe in their own metric tempo as veritable tissues of a thing alive."

Eugene Kolisco: "It is a peculiar fact that all the great astronomers of the 15th and 16th centuries were deeply convinced that the whole universe was a huge living being. Science is on its way to discover that all minerals originate from living things. Life never came into being on Earth, only dead matter has, out of the original life process."

Even during the height of Western culture, the Greeks thought of the Living Planet organism as a fact of life.[6:Hughes]

J. Donald Hughes: "Pythagoreans held the cosmos to be spherical, animate, ensouled, and intelligent."

Guy Murchie: "As early as the sixth century B.C., Greek philosophers like Thales taught that life is a natural property of all matter."

Empedocles: "There is no birth in mortal things and no end in ruinous death. There is only mingling and interchange of parts, and this we call Nature."

Hermes Trismegistus: "The Earth is a living creature, endowed with a body which men can see and an intelligence which men cannot."

Plato: "Whence can a human body have received its soul, if the body of the world does not possess a soul?"

Tyrone Cashman: "For the Greeks, the individual human self was integrated into the unified living cosmos."

Some philosophers believe that the Socratic method of thinking triggered our culture into comparing one aspect of the environment with the other. In using this process, Western people began to perceive themselves as thinkers, comparers, and took on that self-identity rather than remaining a more holistic, feelingful organ of the Planet organism. Greek philosophers abhorred the cyclical arguments needed to understand Nature's cycles, as firmly as they believed that Nature abhorred a vacuum.

Over time, comparative logic and reason led to scientific conceptualization. A major premise of this type of thought is that by stabilizing one factor of life, the motion and properties of other entities may be repeatedly disclosed, compared and measured against this stable control or premise. That process makes taboo the immeasurable, subjective values of sensations and feelings.

The accuracy of scientific thinking with respect to motion (but not emotion) led Copernicus to describe the physical solar system, centering it around the sun rather than the Earth. Further, by using scientific methods, others could affirm the facts of motion he disclosed. The process proved more trustworthy than human frailties, while religiously and politically, it caused an uproar louder than would occur today if someone proved undeniably that the moon really were made of green cheese. Those who used science became as empowered as if they personally owned nuclear weapons today. Scientific thinking became a treasured, profitable mode of thought that made possible exploitation without guilt.[6:Cashman]

By the 1700s, immediate, not long-term, qualifying and quantifying became the accepted means of knowing the environment and the universe. In turn, the Earth became more known for its immediate qualities and quantities rather than as a unified, timeless whole. Leading thinkers like Hobbs, Descartes, Bacon and Newton were rewarded for mentally subdividing the Planet into its mechanical, measurable parts. Separated from time and emotion, these parts were thought to be governed by magic or by God. The Bible was rewritten and first printed during this time period and may reflect the period's outlook.[6:Kubrin] In time, the Living Earth disappeared into a series of repeatable facts and figures, and in the process, Western consciousness lost sight of the Earth's life attributes and integrity.

With respect to Planet Earth, scientific thinking usually overlooks the fact that no aspect of the Earth is stable or constant. Nor can scientists consistently measure sensations and feelings. Thus, few, if any, natural controls exist by which to measure the whole Living Planet. And scientists seldom agree on each other's predictions with respect to their findings about the Earth.

Scientifically, we measured the Earth against our artificially stabilized standards, not Nature's ever-changing reality. And as a rationalization for developing, improving and exploiting the "dead Earth's" resources, scientific inquiry gained religious stature. Against the vehement arguments of Living Earth believers, coal was dug in England out of "Mother Earth's skin," empowering industry. As history denotes, the objectively ritualized, white-coated, esoterically languaged scientist attained priesthood.

Today, that priesthood is being confronted by both itself and its effects.

The Gaia Hypothesis

Lawrence Joseph[10] describes the 1986 laboratory of James E. Lovelock in rural southwestern England. Spectroscopes, radiation detectors and microcomputers sit incongruously in a field. Lovelock, a brilliant, established biologist and inventor uses them to decipher what an army of his adherents in the scientific and environmental communities would argue is one of the Earth's greatest truths.

The British scientist asserts to the *New York Times* and in his book **Gaia: A New Look at Life** that life has direction: "The Biota,— the sum of all living things, including plants, animals and microorganisms—not only profoundly affects the Earth's environment, but acts to maintain and enhance life on the planet."[14]

As suggested by William Golding, Lovelock named this theory the Gaia Hypothesis, after the Greek goddess of the Earth. The Gaia Hypothesis states that what we think of as life, influences the Earth in order to sustain itself, that the Planet itself is really a single, unified, living system, not a conglomeration of disconnected parts and discontinuous functions.[6:Cashman]

As has been noted already, the Gaia theory's insistence that the Earth is a self-controlling, whole system has drawn the interest of scientists worldwide. At its most basic level, the theory forces geologists and biologists to reckon with each other's work while confronting the question: How important is life to the evolution and functioning of the Planet Earth? This, in fact, is the subject of an international conference on the Gaia Hypothesis sponsored by the American Geophysical Union.

Meanwhile, an underscoring of Lovelock's theory occurred in August 1985 at the week-long conference entitled IS THE EARTH A LIVING ORGANISM? That Gaia Hypothesis Symposium, sponsored by the National Audubon Society at the University of Massachusetts,

Facts From Life Science, *Boulder, Utah*

had to turn away more than one hundred prospective speakers and still closed with a proceedings that ran to thirteen hundred pages.

But who is Dr. James Lovelock that his theories can shake the entire scientific community? Well, Dr. Lovelock is a scientific maverick honored by the establishment, one of the very few independent, unaffiliated scientists to become a fellow of the Royal Society, the British counterpart of America's National Academy of Sciences. His interdisciplinary credentials include a Ph.D. in medicine from the University of London, a Chemistry professorship at Baylor College of Medicine in Houston, a Cybernetics professorship at the University of Reading in England and Presidency of the Marine Biological Association of the United Kingdom.

> Lovelock is an impeccable scientist, a prolific inventor who supports his own research with royalties on his many patents and other scientific devices, and retainers from Hewlett-Packard, Shell and other major corporations. His most important invention is the electron capture detector. First developed in 1957, it is still the most sensitive instrument around for analyzing the air. The palm-sized device was used by Rachel Carson to gather the data on pesticides for her 1962 book on environmental hazards, **Silent Spring**. A decade later, Lovelock used his detector to demonstrate that chlorofluorocarbons were accumulating worldwide. This led to recognizing the problem of ozone destruction by fluorocarbons.[10]

Lovelock has the credentials.

Disequilibrium Puzzles
Studying other planets' atmospheres for the National Aeronautics and Space Administration led Lovelock to the Gaia Hypothesis. He reasoned that the gases in a planet's atmosphere should react according to the laws of chemistry and physics to form stable compounds and settle into a lifeless equilibrium like that found on Mars and Venus. However, the Earth's atmosphere defies this natural expecta-

tion. Gases coexist when they should combine, and elements and compounds appear as gases when they should remain settled as solids on the surface. The only explanation for this wild disequilibrium was that the Earth's plants, animals and bacteria continually inject gas and energy into the atmosphere.

Particularly struck by the vast amounts of free oxygen in the terrestrial air, Lovelock wondered why this highly volatile gas did not react with other elements, such as carbon, to form stable compounds like carbon dioxide. Carbon dioxide is a dominant gas in the atmospheres of other planets yet accounts for only three-hundredths of one percent of Earth's atmosphere. Similarly, Lovelock wondered what keeps the oxygen content in the Earth's atmosphere at approximately 21 percent despite extreme differences in the globe's chemical composition. If it shifted a few points higher, fires would burn out of control; at lower than 12 percent, most organisms would die. And there were other questions as well.

The ocean's salt content, 3.5 percent by weight, remains roughly constant while runoff from the continents and submarine volcanism

Rock Solid Evidence, *Diamondville, Wyoming*

dumps more than five hundred megatons of salt into the water every year. Were salinity to climb, when it reached 6 percent, virtually all ocean-dwelling organisms would die. Why has that not happened?

The Earth's average surface temperature has remained relatively constant between 50 and 68 degrees Fahrenheit. Yet over the course of approximately 3.5 billion years of life on Earth, the sun's output of energy may have increased by as much as 30 percent. Just a 2 percent increase in solar output would ordinarily cause the seas to boil, and an average temperature rise or fall of just 2 degrees would prohibit life. Lovelock claims these life-terminating changes are prevented by planetary temperature controls such as albedo, the Earth's reflective surface which changes with increased or decreased cloud, snow and vegetation cover. Regulation of atmospheric carbon dioxide content also regulates the Planet's temperature because carbon dioxide prevents long-wave heat escape into space through what is known as the greenhouse effect. How different from Venus and Mars, our neighboring planets, which both increase their temperature as the sun emits more energy, though Venus is closer and Mars more distant from the sun than is the Earth.

One can only conclude that the Earth plays a self-regulatory, self-organized role in determining its own temperature. Ordinarily in Nature we only observe this ability in living organisms.

Lovelock concludes that the climatic and chemical properties of the Earth seem to have always been optimal for life. Unlike those on other planets, Earth's waters neither freeze nor boil away. For this to have happened by chance is virtually impossible.[14]

A traditional argument against Gaian thinking is that the Earth is the best possible environment for the organisms that have adapted to it. This concept neglects the fact that every organism influences and impacts its environment, that the air, ocean and rocks are made or changed by living organisms. The conventional argument also exists that a giant planning body would be necessary to govern and guide Earth. This "higher authority" reasoning dismisses the possibility that organisms sensitively acting in concert can determine their own destiny and direction through self-regulatory processes.

The Life-Friendly Atmosphere

Lewis Thomas notes that: "It takes a membrane to make sense out of disorder in biology. Energy must be caught, held, stored and released in precise amounts. Each cell and its structures do this with energy flow—all of which originates with the sun. Life means holding out against equilibrium—banking against entropy by using membranes. The Earth has constructed its own membrane, the atmosphere, in order to edit the sun."[24] Thus, as might any living organism for survival, the Earth acts to reduce its increasing temperature.

The salinity, the albedo and greenhouse processes are subject to the membrane-like conduct of plant and animal life. For example, it has been noted by satellite that when warm air passes over the ocean, the water turns white from quickly reproducing microorganisms in the sea. These organisms remove heat-retaining carbon dioxide from the warm air mass and combine it with calcium in the sea. The resulting calcium carbonate composes the shells of these sea creatures. They soon die and fall to the ocean bottom where they accumulate as limestone and oil, a carbon sink.[6:Lovelock] Removing carbon dioxide from the atmosphere by this method reduces the greenhouse effect, thereby lowering temperatures.

The calcium in the sea from which microorganisms make their calcium carbonate shells comes from the wearing down and weathering of calcium silicate rocks on the land. Living organisms such as soil bacteria, worms and trees help to weather these rocks, which in addition to calcium also form nutrients and soil. The latter nurture more organisms whose respiratory processes release carbon dioxide as carbonic acid into the soil. This acidity in turn hastens rock weathering. The organism-controlled calcium salts thus set free are soluable and eventually wash into the sea. There, sea creatures combine the calcium with carbon dioxide, make their calcium carbonate shells and form ocean bottom limestone sediments upon their death.[15]

In addition, scientists have found that limestone's dense sea deposits may trigger and regulate the mid-oceanic rift activities that cause sea floor spreading and continental drift. As a result, the relative position of the continents and seas is thought to partially determine the Planet's ambient temperature and salinity.

In the December 18, 1986 issue of *New Scientist*, Lovelock notes, "The net effect of this complex set of interactions is an increase in the biomass, the pumping of carbon dioxide from the air, and a cooler climate. For a healthy Planet, the balance is reached with a climate that is cooler than optimum."[15] It is as if the Planet is a living embryo, surviving by continually adapting to the increasing temperature of its solar womb.

In all these ways, previously considered "dead" physical forces, like limestone geology, regulate and participate in the life of Earth, just as hair and teeth do for us. Even little bugs help run the Planet.

Not accidentally, the nine major gases in our atmosphere are maintained in exact vital proportions which have not varied in millions of years. This delicate balance can only be maintained if the Earth itself monitors these gases and directs the terrestrial activities that produce them. The gases in turn maintain life. For example:

> nitrogen maintains atmospheric pressure
> oxygen energizes all chemical reactions
> carbon dioxide feeds photosynthesis and limestone and insulates the Earth
> methane regulates oxygen and rids mud and silts of natural toxins
> nitrous oxide regulates oxygen and ozone
> ammonia regulates the Earth's acid-base pH
> sulphur gases participate in the sulphur cycles
> methyl chloride regulates ozone
> methyl iodide transports iodine[6:Lovelock]

Like an incompatible blood transfusion into a person, an imbalance of these gases would prove catastrophic for global life. The Gaia Hypothesis suggests that this doesn't happen because life itself maintains their proportions. This may be seen by comparing the gas ratios of non-life planets to those of Earth.[6:Lovelock]

Comparison of Gases, Pressure, Acidity and Temperature on Planets With and Without Life [22]

	Without life	With life
Carbon Dioxide	98%	.03%
Nitrogen	1.9%	79%
Oxygen	trace	21%
Argon	0.1%	1%
Average Temp. (C)	240-340	13
Pressure (bars)	60	1
Average Acidity	highly acidic	neutral

The same drastic differences occur when scientists compare the composition of the oceans and the air of our world to that of a hypothetical, chemical equilibrium world.[14] They calculate that all the carbon buried in oil, coal and limestone deposits equals that which would combine with atmospheric oxygen to form carbon dioxide. They postulate that, as with other planets, the Earth once had a carbon dioxide atmosphere. Like a living organism, the Earth's plant and animal organic system removed carbon from the air. By regulating the Planet's geology, the Earth's life force buried the carbon producing the life-giving free oxygen in the Earth's atmosphere. Today we release carbon into the atmosphere as we burn coal and oil. And our acid rain releases it from limestone. Thus our modern lives are changing the atmosphere's composition and temperature, the very things that are needed for life.

Recent findings indicate that the Planet can no longer bury carbon as fast as our industrial society produces it as a waste product. Today, scientists note atmospheric carbon dioxide rising and the Planet warming. The predicted long-term effects of these changes due to the modern way of life include the triggering of a new glacial age, heat waves, drought, weather extremes, desertification, and coastal flooding from sea expansion or other readjustment of the Planet organism's metabolism to our discomfort or demise. In fact, it has already begun. The Planet's temperature has risen 1/2 a degree Centigrade in modern times, leading to drought and famine crises in Africa.

Similarly, scientists report that the cause of Chesapeake Bay's demise is no single industry or pollutant. Instead, they strongly suggest that it is our total modern way of life that is causing the Bay's destruction.

The homeostatic phenomenon governing levels of atmospheric carbon dioxide also appears to operate with respect to the Planet's acidity. Presently the global biosphere produces approximately one thousand megatons of ammonia each year—almost exactly the quantity needed to neutralize the strong sulfuric and nitric acids produced by the natural oxidation of sulphur and nitrogen compounds. To date the Planet has not adapted to these industrial exhaust compounds' production of acid rain, nor may it ever.[6:Lovelock]

Learning From Experts, *Archibald Biological Laboratory, Florida*

Without Gaia's influence, atmospheric nitrogen would soon be oxidized by lightning discharges and would wind up in the sea as nitrate ions. Instead, it may also exist in the atmosphere as a gas, placed there by "little bugs," such as de-nitrifying bacteria and cell metabolism.[6:Lovelock]

Maintaining Life-Friendly Temperatures

There is more to the biotic community's temperature regulation of our Planet than just what is mentioned in the preceding paragraphs. For example, the *New York Times* reports that Lovelock's discovery of dimethyl sulfide (DMS) in ocean water suggested to him that the compound may also exist as a gas. Lovelock's speculation was subsequently verified by Prof. M.O. Andreae, Florida State University at Tallahassee, who suspects that DMS is necessary for cloud formation over the open seas, clouds that are essential to what Lovelock calls the "planetary refrigeration system." Clouds also become thunderstorms and rain whose lightning may produce the atmospheric ozone which shields land life from harmful rays. But DMS is produced by microscopic marine algae. Andreae says that if they didn't make DMS, the Earth would be much warmer. Life and geology become identical. Here again, little bugs help run the Planet.

The possibility that ocean-dwelling organisms are helping to regulate the Earth's temperature and ozone layer elevates the Gaia Hypothesis to a status in opposition to the traditional Darwinian belief in evolution-by-competition. Darwinian theory doesn't see any coupling between the evolution of the environment and that of living organisms. What Gaia sees is a tightly coupled process where the evolution of life and of rocks, oceans and atmosphere are so closely joined together that they are really one single process. Lovelock is writing a new book which further documents local and global examples of this process in action.

The Role of Microorganisms

Natural selection does not take place in a neutral environment, because the environment is not a constant. For example, 100 percent of living organisms interrelate and give off gas 100 percent of the time, constantly changing the atmospheric environment. Termite intestinal bacterial activity alone can produce enough methane gas to dramatically regulate the oxygen/carbon dioxide content of the global atmosphere —more bugs doing life work.[31] And human dependence upon intestinal bacteria for digesting food and making Vitamin B12, not to mention our amoeba-like white blood cells that control disease, indicates that many microorganisms regulate our personal biology as well as the Planet's.

We're seldom aware that during life's first two billion years, which scientists suggest represents two-thirds of the history of life on Earth, bacteria invented all of life's essential chemical systems, developed fermentation, photosynthesis, oxygen breathing and the removal of nitrogen gas from the air. The descendants of bacteria that swam in primeval seas, breathing oxygen three billion years ago, exist now in all plant and animal bodies as mitochondria, without which neither plants nor animals can utilize oxygen and live. Mitochondria have their own DNA genes and an independent reproductive cycle from the cells they inhabit.[16]

The cooperative living relationships, symbiosis, between bacteria and their environment gave rise to recent life's organs and organ systems. In other words, we are recombinations of powerful bacterial communities with a multibillion-year-old, cooperative, networking history. Microorganism activity metabolizes the Planet similarly to our intestinal enzymes digesting food into nutrients for cell life.

Simple bacteria, those ancient organisms and forebears of the Earth's biotic substrate, are today capable of expanding and altering themselves and the rest of life, should we annihilate ourselves. Their role in a living organism is exemplified in our body, which consists of ten quadrillion animal cells and another one hundred quadrillion bacterial cells.[16] And, when we die, bacteria recycle us into the global ecosystem.

The integrated, symbiotic life picture is found to exist in all the sciences. Even subatomic particles have been found to exhibit consciousness, and light is capable of changing its form from particles to waves.[27] Predictably, scientists will yet discover that through microorganism activity, the ratios of sea salts and soil chemicals coordinate with Gaia's global metabolism and its ongoing relationship with the sun.

The Rocks Are Alive
Planet Earth can't be just dead rock, because over time a rock's "dead" orogeny and geological cycle are not lifeless. With patience, one can notice almost any stone's signs of life. Its cycles are more like those of a person's teeth or fingernails, continually growing and recycling as they wear away.

We now know that just as living organisms' seashells become rock, which we call limestone, some portion of each rock's erosion and rebuilding includes one or more life-form activities. Three billion years ago, growing microbes trapped minerals to form layers of rock in shallow oceans, yet they were but imitating a process already known to rocks.[16] Clay particles and crystals have clearly demonstrated similar lifelike growth properties.

All Points of View, *Arches National Monument, Utah*

Crystal growth amounts to elementary procreation, another basic attribute of life, for each species of crystal can accept only its own kind of molecules to grow by accretion. Indeed, each molecule, in order to be accepted, must attach itself at exactly the correct angle for whatever substance it is: 90 degrees in the case of salt or silver; 60 degrees in snow or quartz; odder angles in odd crystals like axinite or rhodonite. This is a genetic process because the crystal lattices themselves serve as "genes." They only admit one specific kind of molecule to fasten and grow upon them. This was no doubt life's first and simplest reproductive technique on Earth a few billion years ago. From it have evolved dozens of different, apparently organic forms, such as the pealike clusters of bauxite crystal, the hairy ones of asbestos, the sea-urchin-shaped radial globes of warelite, the "asparagus sprouts" of limonite, the foliated nuggets of copper, the nervelike branches of pyrolusite, and the serrate leaves of muscovite. All of these are crystal species with growth habits that dramatically reveal their kinship with the rest of us.[17]

The additional fact that very hard stones such as whewellite, found in coal, weddellite, discovered in Antarctica, and struvite, magnesium ammonium phosphate hexahydrate, all commonly grow from "seeds" in human kidneys and bladders, building themselves up layer by layer like their crystal counterparts outside, is just further evidence that rocks can be a very intimate part of life—even your life.[17]

Viruses epitomize rock life. They "inhabit" all organisms from bacteria to whales and are now proven to be inert crystals when dormant; yet when the right amounts of moisture and warmth awaken them from their stony slumber, they spring eagerly to life, invade other beings, reproduce themselves and evidently feel utterly at home in each and all of the kingdoms.[17]

With training, it's easy to recognize rocks as basking, shivering, living beings. Scientists conclude that some rocks grow, un-grow, glow, radiate, disintegrate, get ill and regenerate.[17] **Saulk Laboratory's Leslie Orgel has recently discovered a DNA-like molecule**

that formed spontaneously from simple carbon compounds and lead salts—in the total absence of living cells!

From the moon, astronauts can view the inert, rocky Planet Earth as a

> superorganism basking in the nourishing environs of her paternal star, the sun. The nourishment is the "solar winds" atoms, molecules, heat, light and other radiation which continuously stream out of the sun. And like smaller creatures, the Earth literally stirs in her "sleep," while quietly breathing in her vegetal way, her skin slowly wrinkling, her sore spots volcanically breaking out but almost as quickly healing over again, her magma juices circulating, her lithic flesh, skin, and breath metabolizing, her electromagnetic nerves flashing their increasingly vital messages. Intermittently she also rumbles with confined internal gases, utters earthquakes, itches a little, dreams strange dreams, and through her inhabitants, feels self- conscious.[17]

And when internal magmas and core minerals flow, the Planet's magnetic lines of force change their position and consistency, and the magnetic poles shift. In turn, certain plants and animals sense these changes and redirect their behaviors. This again suggests that the Planet's geology and organisms play a role in evolution and in life's survival. Endless examples of this exist.

In 1977, two Canadian psychologists wrote a book called **Space-Time Transients.** They looked at the relationships between geological stress and faulting areas, and mental processes that can occur. The basic hypothesis was that a person's consciousness is based on very subtle electrical patterns in the brain. These are normally systematically balanced and undisturbed by environmental fields. At certain times, however, transient electrical fields from geologic phenomena, like electrical energy from fault scarp friction or lightning, can upset the electrical field of the brain. The hippocampus, known as the horns of Ammon portion of the brain, is most electrically sensitive or unstable. It is a dual, horn-shaped structure in the mid-brain limbic system that is tied to the emotional and short-term memory part of the brain. **When the hippocampus is disturbed electrically, the whole way we experience reality is distorted and we begin to see in strange ways. Our "rock-solid" memories are modified by the landscape!**[6:Steele] This may influence our relationships and programmed consciousness.

Penfield's probes of the hippocampus gave the greatest rise to hallucinations; he showed that a disturbed hippocampus creates numerous inappropriate images. A spray of images from different periods can occur, as may a confabulation of images that never existed. Thus it appears that sub-surface geological stresses can produce frightening and terrifying mental images.[6:Steele]

Archaeologists note that throughout the ages, certain geographical sites have properties that make them consistently chosen for sacred or religious purposes. They hypothesize that these natural areas contain geologic waves and forces that influence the psyche and call attention to Gaia's activities. Modern persons who solo or meditate for long periods in these areas gain notable strengths, peace of mind and a bright, burning look in their eyes.[6:Milton]

Lyall Watson says it is no longer possible to deny that our minds and thoughts influence our environment, while the most recent cosmologies all include consciousness as an active reality factor. Most interesting, however, is the fact that the new concepts of how the world works are similar to the age-old beliefs of non-literate people everywhere.

Jeff Goodman, a psychic archaeologist, contends that human behavior, social activity and thoughts play an important role in the occurrence of earthquakes and climatic changes.[6:Steele]

Dr. John Steele of the Dragon Project says, "Cultures used to build structures on principles of geomancy—places of power, energy sources and sinks in the landscape, places that feel (body-knowing) attractive. I noticed in Ireland, stones in the middle of fields called fairy stones. I was told by locals that if you moved the stones, the entire productivity of the field diminishes: cows don't give milk,

The Writing on the Wall, *Grand Gulch, Utah*

chickens don't lay eggs, the farm goes downhill."[6:Steele] Obviously, "inert rocks and our inert planet of rock" display the properties of life.

Jim Lovelock points out that it is no longer sufficient to say, as did Darwin, that "organisms better adapted than others are more likely to have offspring." It is necessary to add that the growth of an organism affects its physical and chemical environment and that, therefore, the evolution of the species and the evolution of rocks are tightly coupled as a single, indivisible process.

Life in Waves and Electrons
Life and awareness qualities are even found in electrons and other subatomic particles, basic components of every atom.[27] Never sleeping, they produce, react to and resonate to magnetic fields. Scientists observe them "consciously" choosing their own path and rotating a quadrillion times per second with their own gyroscopic bias, like the "torque of mind" of a top. Electrons can travel the Earth's circumference five times in a second, yet while orbiting an atom's minute nucleus at that speed, they know what other electrons are doing and act accordingly. Although distanced from their polar opposite, **they appear conscious of it and reverse themselves AS ONE when their polar partners are reversed!**

Meanwhile, scientists report observing similar behavior in South Pacific monkey populations. Monkeys isolated from each other learn identical behaviors without communicating. Isolated laboratory rats have also been reported to learn this way from each other.

Robert Sheldrake hypothesizes that the universe governs itself, and that the characteristic forms taken up by molecules, crystals, cells, tissues, organs and organisms are shaped and maintained by specific fields akin to gravity and electromagnetism. These he calls morphogenetic fields.

Lodestone, the magnetic "rock that loves," is the result of electron activity in its iron particles.[17]

"The wave aspect of the atom is the mind aspect of matter," states Niels Bohr. Interestingly, the atoms that make up protoplasm, carbon, hydrogen, nitrogen, oxygen, phosphorous and sulphur are in turn the atoms which have the highest wave activity. They entertain the greatest variety of combinations so necessary to organic life.[18]

Networks, Sensations and Consciousness
Lovelock concludes that, like a thermostat, the Earth is "homeostatic" and follows what Walter B. Cannon called the wisdom of the body. The biosphere, geography and minerals together function as a self-regulating living organism, coordinating their vital systems to compensate for environmental changes and threats. This intimates at least a rudimentary form of shared sentient, global communication and intelligence.[6:Young] Recently, individual white blood cells have been observed to communicate with each other when they touch.

Modern ecological science documents the interconnection of all matter and entities. It supports the poet who said, "When you pick a flower, you trouble a star." Gaian science recognizes global interconnectedness as a self-organized, regenerative global life system. The networking of all entities interacts as would the flesh and blood of a living organism. [6:Young]

Lovelock states that life is a phenomenon that exists on a planetary scale and that there can be no partial occupation of a planet by living organisms. Such a planet would be as impermanent as half an animal. Living organisms have to regenerate their planet; otherwise, the irresistable forces of physical and chemical evolution would soon render it uninhabitable.[15]

The Earth behaves spontaneously. Its self-organized, integrated acts speak clearly of life sensations and reactions. Modern people doubt the Earth organism's sentient nature simply because they don't inherit sensitivities to it, don't observe it or through their upbringing have lost the ability to experience many of the Earth's emotion-like interactions. Yet scientists have located over thirty basic sensations of the Planet's networked plant, animal and mineral nervous systems, not including sense of mind, such as that which ants display when they place mud across a sticky barrier in order to cross it. The following list is from **The Seven Mysteries of Life** of Life by Guy Murchie and is reprinted with permission of Houghton Mifflin Company. (Copyright 1978 by Guy Murchie.)

PLANET SENSATIONS [18]

The Radiation Senses

1. Sight of light, polarized light and seeing without eyes, such as the heliotropism or sun sense of plants.

2. The sense of awareness of one's own visibility or invisibility and consequent camouflaging.

3. Sensitivity to radiation other than visible light, including radio waves, x-rays, gamma rays, etc.

4. Temperature sense, including ability to insulate, hibernate, and winter sleep.

5. Electromagnetic sense and polarity, which includes the ability to generate current.

The Feeling Senses

6. Hearing, including sonar and ultrasonic frequencies.

7. Awareness of pressure, particularly underground and underwater.

8. Feel, particularly touch on the skin.
9. The sense of weight and balance.
10. Space or proximity sense.
11. Coriolis sense, or awareness of effects of the rotation of the Earth.

The Chemical Senses
12. Smell, with and beyond the nose.
13. Taste, with and beyond the tongue or mouth.
14. Appetite, hunger and the urge to hunt, kill or otherwise obtain food.
15. Humidity sense, including thirst, evaporation control and the acumen to find water or evade a flood.
15a. Hormonal senses, like pheromones and other chemicals to which plants and animals respond.(MC)

The Mental Senses
16. Pain, external, internal, mental or spiritual distress.
17. The sense of fear, the dread of injury or death, or of an attack.
18. The procreative urge, which includes sex awareness, courting (perhaps involving love), mating, nesting, brooding, parturition, maternity, paternity and raising the young.
19. The sense of play, sport, humor, pleasure and laughter.
20. Time sense and, most specifically, the so-called biological clock.
21. Navigation sense, including the detailed awareness of land and seascapes, of the positions of the sun, moon and stars, of time, of electromagnetic fields.
22. Domineering and territorial sense.
23. Colonizing sense, including the receptive awareness of one's fellow creatures, sometimes to the degree of being absorbed into a superorganism.
24. Horticultural sense and the ability to cultivate crops, as is done by ants who grow fungus, or by fungus who farm algae.
25. Language and articulation sense, used to express feelings and

convey information in every medium from the bees' dance to human literature.

26. Reasoning, including memory and the capacity for logic and science.

27. Intuition or subconscious deduction.

28. Aesthetic sense, including creativity and appreciation of music, literature and drama.

29. Psychic capacity, such as foreknowledge, clairvoyance, clairaudience, psychokinesis, astral projection and possibly certain animal instincts and plant sensitivities.

30. Hypnotic power: the capacity to hypnotize other creatures.

31. Relaxation and sleep, including dreaming, meditation, brain wave awareness and pupation, which involves cocoon building and metamorphosis.

The Spiritual Senses

32. Spiritual sense, including conscience, capacity for sublime love, ecstasy, a sense of sin, profound sorrow, and sacrifice.

Many non-Western cultures show sensitivity to these aspects of the Living Earth. Perhaps collectively these sensations make up Teilhard de Chardin's noosphere, a consciousness grid that encompasses the Earth. We may feel these sensations unconsciously in beautiful wild places, as did people like John Burroughs, John Muir and Aldo Leopold. Perhaps these sensations led Emerson to write, "The Earth laughs in flowers," and Thoreau to note, "Wilderness is a civilization other than our own."

Mainstream science's mechanical objectivity often denies as unscientific, ignores or rejects the findings of research that delves into the full scope of sensations. New sciences such as geopsychic phenomena, geomancy, psychology, extra-sensory perception, geophysiology, psychokinesis and telepathy may be detecting sensitivities that some individuals have not fully lost or that ancient peoples enjoyed.

The Legacy of Life Science
A vital, often overlooked phenomenon of physical science research may ultimately assist scientists' acceptance of Gaia. Although physicists recognize that the nature of particles and of waves is different, when they use particle-measuring instruments to determine the nature of light, light behaves as if it consists of particles. But when physicists use wave-measuring instruments on light, light behaves as if it is a wave. Yet, it can't be both at the same time, for that would be like saying black is white.[27] Thus, **the pre-existing expectations or bias of scientists appears to enter into their research and into the nature of reality!**

This expectation factor probably influences scientific considerations of Gaia as well, for scientists have yet to consider magnetism, gravity, electricity and other physical forces as manifestations of a single universal emotion, affinity, the love of surviving by building stabilizing relationships.[26] Yet sensations and emotions, like affinity, are unique properties of life. Excluding them from research can produce blindness with respect to the Earth being a living organism.

The traditional relationship of science and the media toward Gaia reflects our general mistrust of the universe's underlying pulsating affinity desire, the life force that sparks Nature's self-organizing wisdom. For although scientific methods place organisms with common ancestry in the same family, our habitual anti-Nature bias prevents us from validating the kinship between people and their living mother Planet. Our dualistic thinking rejects the nurturing, human mother-child relationship as being a microcosm reflecting the Earth-person continuum. Yet, practically nil is the scientific probability of an infant's nursing chemistry resulting from laws of chance alone. Not nil are the millions of dollars spent to scientifically validate what a pregnant mother's body and natural senses automatically produce.

Gaia bridges Planet-person duality. It sees the enveloping atmosphere and underlying geology as a single unit, as cooperating parts of the global life system. If the Planet is a coherent, self-sensing entity, we are circumscribed by this entity. If Gaia exists, then, like a fish in the water or an infant in the womb, we are inside Gaia.[6:Abram]

No matter their claims for objectivity, most scientific inquiries are tainted by our culture's "conquer Nature" bias. And so discoveries like Gaia, which validate Nature's ways, take decades for acceptance while we almost immediately acclaim, fund and utilize research that controls or destroys Nature—as witness the thousand new chemicals we introduce each year whose harmful effects are still unknown.

Gaia shifts the locus of creativity from the human intellect to the enveloping world itself. The creation of meaning, value and purpose no longer hovers inside the human physiology, for it already abounds in the surrounding landscape.[6:Abram]

Human perception is a constant communion between ourselves and the living world that embraces us. Artificial laboratory situations and instruments have misled scientists into conceptualizing perception and knowledge as a physically passive, internal, cerebral or mechanical event. As with other life forms, however, our senses never outstrip the conditions of the living world, for they are the very embodiment of these conditions.[6:Abram] The Gaia Hypothesis challenges "dead Earth" scientists, politicians, environmentalists, economists and educators to take this fact into consideration.

The Earth Lives! Know it to be true!

One Breath Have All, *Annieopaquotch Mountains, Newfoundland*

DRAWING CONCLUSIONS

Scientific thinking suggests that Earth is a living organism, much of whose activity deals with editing the Sun's heat and rays, to life's benefit. We diagram the Earth organism in conjunction with the Sun that both empowers and threatens it.

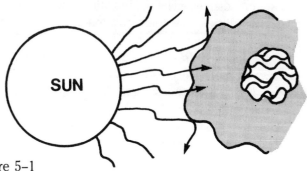

Figure 5-1

STUDY GUIDE—CHAPTER 5

The author collects scientific information about the Living Earth Theory. Experiment with the following exercises and locate or design others. Use them to create a personal science for yourself that helps you internally and externally recognize Planet Earth in new ways.

1) Practice writing subjectively. Select an environment(s) of interest to you. In writing, describe that environment by how you sense and feel it. Don't let yourself describe it objectively; show your emotional reactions to it. Attempt to locate your feelings and validate them.

2) Purposely disconnect from mediating the natural environment.

a) When in a natural area, repeat the word "one" rhythmically. This prevents your mind from creating other words. It also properly labels your relationship with Gaia. Note if and how this exercise modifies your experience in the area. Try this same experiment with *other Gaia-compatible messages such as "I'm alive, you're alive."*

b) Match colors that you see around you with the same colors you might see or feel within you.

c) For a minute, imagine an eggbeater scrambling your mind inside your head. Turn off the image and experience the natural world for a few moments before your mediation powers again come into play.

d) Find a musical note or notes that express how you feel while in a natural area. Hum or sing them.

e) Listen to sounds about you and hum in resonance with them.

3) To relax anytime, turn off your red thoughts and feelings by counting backward from one hundred while visualizing or drawing each number in your mind before going on to the next one. This is an excellent way to go to sleep, so don't do it while you're driving.

a) Relaxation may also be accomplished by visualizing a natural area, like the seashore.

4) Research or discover other Gaia connectors that work for you. See if they work for others.

NOTES

On this page or elsewhere, write down any thoughts or feelings that have come to you from reading this and previous chapters.

Purifying Misconceptions, *Acadia National Park, Maine*

CHAPTER SIX
VOICE OF A LIFE SYSTEM

Although a good, working definition of life and death has never been established, when dead things act as if they're alive, we're perturbed. It's like the woman who cremated her husband's body and had his ashes put in an urn which was placed on the piano in the living room. Unknowingly, her guests got into the habit of tapping their cigarettes into it. One day, she frantically called a close friend saying, "I know you're going to think I'm crazy, but I swear Sam's putting on weight."

My Grand Canyon encounter stimulated a Living Earth consciousness that threw new light on every experience that followed. It offered me a new lens with which to view the world. Like a microscope, the lens brought to awareness phenomena right before my eyes that I could not see through modern society's "dead planet" lens.

You, too, use limiting lenses, as we discussed in Chapter 1. For example, as you now read this page, out of habit you pay attention only to the words and phrases upon it; you are not conscious of the individual letters. Yet, once confronted with this phenomenon, you not only become aware of the individual letters, you realize you didn't previously perceive them. The confrontation offers you a new lens. Now you can see either the words or the individual letters, at your discretion.

Yet, there is still another thing to which you are blind as you read this page, and again, you must develop a different lens to see it. For although you now see these words, letters and pages, you don't see the air between you and this page. Air is no more invisible than the individual letters you couldn't see previously; you simply must use the correct lens to perceive it. And since air is so extremely important to life, we can't afford to overlook it.

On the Expedition, the Living Earth frame of reference made me note that I take air for granted. This first happened on a short solo in Newfoundland's Gros Morne National Park. A solo usually consists of spending an hour or two in an inviting natural place. We use solos to give our consciousness space to integrate sensations or thoughts arising from visiting new areas.

Plagued by the everyday problems that fill my mind, I sit alone high on a rock face overlooking the grandeur of Western Brook Fjord. I wait to see what happens. During these periods, everything that enters one's head is considered a fact of that time and place, including the mosquito that flies up your nose.

As I admire the spectacular array of woodland, cliffs and lakes, the distractions of daily life begin to fade. Soon a gentle breeze alerts me to that which I've never before considered: I ignore the air. I look through the air, not at it. I overlook Nature's lifeblood, the atmosphere. In retribution I attempt to sense this. Closing my eyes, I listen to myself breathe, savoring each life-preserving exchange between myself and the Planet. This conscious connection with the natural world relaxes me. It helps me catch up with myself. For the moment, Nature's forces erase the tensions arising from my memory's expectations and conflicts with the past and future.

After a while, for a reason unknown, I purposely stop breathing. Does the Planet signal me to do this? Why should I think to do this now? During my hundreds of previous solos, I have never entertained this idea.

I continue to refrain from breathing and disconnect myself from the substance of the Planet. It and I no longer relate through respiratory atmospheric contact. Temporarily, I break an ancient, biologically inherited relationship that established itself over 400 million years ago when creatures first lived on land. The respiration bond has, over those aeons, repeatedly affirmed its validity for sustaining life on the land. I threaten it when I stop breathing. It, in turn, responds.

Gradually I feel a **tension build** that I didn't sense moments before. I feel my body cry for air. The cry calls from part of me being deprived of another part of me. I am being torn apart. I can feel it happening.

Finally, I can torture myself no longer. My Earth-inherited survival feelings demand that I breathe again and I do. The air, part of the body and soul of Mother Earth, embraces my life. The **release of tension** feels wonderful. That special joy reunites the Earth and me. It confirms that I have re-entered the vital love affair that has sustained us for so long.

In this Nature-governed relationship, the Earth calls the signals. Breathing is not my decision. If I decide not to breathe, I pass out, at which point the Earth takes over, reviving me by giving me natural respiration. Even when I sleep, Nature keeps up my respiration, keeping me alive.

The encounter helps me to sense great truths. It makes me aware that to sustain my life, the Planet communicates erotically through some level of immediate tension-producing and tension-relaxing suffocation feelings or sensations (T-R). T-R has so sustained life since its beginnings. My life was in some way so formed. T-R underlies life's ever-changing pulse and fluctuations. Scientists and psychologists agree with this notion. They recognize that Nature, not people, invented feelings and sensations. Yes, we modify them, or attach them to objects, situations, symbols and images. But invent them? Never! The Planet did that.

Without words, numbers or a cost-benefit analysis, the Earth conveys information through T-R feelings such as hunger, thirst and the sex drive. Its message lets me "see" the air without a visual signal. It tells me to breathe and share my life for our mutual existence.

The Expedition members return from their solos and relate their experiences. I tell of my **inspiring** breathing adventure and learn from one of the students that the words **respirate** and **inspire** derive from the same word: spirit. In other words, **our breathing is a spiritual relationship with the Planet.** If we break the relationship, we expire. Death is Nature's way of telling us to be careful of Nature's ways.

My breathtaking solo gives me still another lens with which to view the natural world. It suggests that the pulsating T-R build-up and reduction cycle of tensions signals my body as to how and when to act. That **now** relationship sustains itself **but fluctuates with T-R.** Similarly, T-R governs our hunger, thirst, sexual needs, temperature, salinity, etc.

Critical observations disclose that T-R can be found in every aspect of Nature. For example, the tension build-up of rabbit populations is released by coyote predation. The coyote, in turn, is governed by the rabbit population fluctuations.

If we place the T-R lens over our eyes, we observe that T-R exists in the orbits of planets around the sun and electrons around an atom's nucleus. We find T-R in the wind, in waves, in weather patterns, in rocks and in molecular relationships.

Breathtaking Inspir(it)ation, *Acadia National Park*

The T-R lens reveals **T-R as a vital part of Nature, a communicating and controlling, constantly pulsating force at the heart of every local or global life system.** Each entity of the universe appears to utilize and react to the immediate vibratory joyful calls and stresses of T-R from its environment.

In many, if not all, entities, T-R translates into feelings, sensations, memory or some other signaling energy resembling consciousness. **Like any other organism, the global life system uses T-R signals to organize, regulate, sustain and regenerate itself on all levels.** T-R is Nature's voice.

Perhaps the most startling realization is that in humans, T-R is experienced as feelings: our sensations, relaxations, stresses and tensions. Human T-R survival feelings come from the Planet's physiology and metabolism and translate into every kind of feeling and sensation we know. They underlie our aesthetics, tastes, fantasies, imagination and emotions. Sensations are the Planet's callings inside us. It embraces us in its emotions. We bathe in them. Even bad feelings, horrible dreams and antisocial fantasies only express the stressed state of the Planet's life within us.

Our feelings are the Earth's way of letting us know what the connection is between the Planet's life and our own. Feelings expand our knowledge of the world far beyond the conventional five senses that we've learned to accept as our sensation limits.

There's another aspect of Nature to be learned from T-R experiences. As you lie on the beach in the warmth and light of the sun, you sense the sun's presence. You know it exists because you feel it. That alone makes it a real fact. You can also know that fact by photographing the sun or by measuring its light with a mind-boggling array of scientific instruments. Although they, too, can tell you that the sun exists without you ever directly sensing it, they can give you neither its warmth nor its natural brilliance.

From the scientific process, we can know that the sun is a hydrogen fusion nuclear reaction, and as a dying star, it has gotten 30 percent hotter during the past three billion years. We can learn that the sun's light takes eight minutes to travel the 93 million miles across the solar system to reach the Earth. In other words, should the sun be extinguished, it would take us on Earth eight minutes to find out. Yet, ignorant of these modern facts, but simply touched by the light and heat of the sun, people and Earth have lived together successfully for two million years or more.

Many of us believe that the sun exists not so much because we sense it, but because scientists say it does. As we demean the light of our sensations, we depend upon knowing life artificially through measurement and machines. From this we lose many of Nature's guiding T-R signals.

If the Earth didn't receive sunlight, most, if not all, of life as we know it would cease, since most life depends upon the availability of oxygen, a gas produced by plants only when sunlight is present. In this respect, Planet Earth is like a giant plant cell T-R that breathes in sunlight and exhales oxygen. Yet today we seldom acknowledge that **the Earth has a respiratory relationship** with the sun; the

Tension-Relaxation: Global Pulse Waves, *Quoddy Head, Maine*

Planet **inspires** an energy gift from the sun and **expires** a life-giving gas: oxygen.

"Respiration" is a scientific term describing photosynthesis and metabolism relationships through the mechanics of chemistry and light and heat physics. But as we can see, the word means **"Re-Spiriting."** Clearly then, Planet Earth has a respiratory-spiritual relationship with the Sun, a relationship that sustains life as we know it. Without it, life loses its spirit and expires. Scientifically speaking, therefore, life is spiritual, and when we allow ourselves, we can experience positive "spiritual" feelings for and from all the entities in our global life community.

You can experience how your spirit thrives on its connection to the Sun. Hold your arm out straight, concentrate on it being strong and unbendable, and then ask a friend to attempt to bend it at the elbow. Resist your friend's efforts as much as possible. Now think about the Sun's energy entering your body through the Earth, streaming through you, through your outstretched arm, and attaching to the environment out beyond the horizon. Again, holding your arm firm, have your friend once more try to bend your arm at the elbow. Does it bend as easily, or at all? Don't you feel much stronger this second time as you recognize yourself as connected to the Earth? That's the spirit!

T-R feelings taught me that they and Nature are immediate. Like Nature, they exist only in the present, in the NOW of life. They tell me that our Living Planet constantly relates through active fluctuating contact like my breathing—like the T-R point and moment where two turning gears intermesh or a wave hits the shore.

Nature is the pulsating NOW of all the entities in the universe physically relating to the moment. They constantly reshape their relationships from the T-R signals and forces they give and receive during the moment. If we remove the static nouns by which we label entities, we discover that underneath the label each entity is actually a **verb**—a communion of pulsating T-R relationships. For this reason we may describe Planet Earth as a **global communion of entities.**

Consciousness of the past and future as well as scientific data can help us understand how Nature works, but they're not Nature. Nature acts. Anything else is artificial.

We create our troubles mainly by viewing and critiquing Nature with data from the past and future, rather than treasuring its moments of sunlight. You can't know a wild river's nature only by studying a graduated flask of its water, nor by demeaning your feelings about its enchantment. Nature is NOW, and that includes the river's response to you as well as yours to it. If we only recognize the river economically, we pollute both our relationship and it.

As it spoke to me in Grand Canyon, Nature speaks to us through gut level and aesthetic T-R feelings and sensations. **Although Earth is illiterate, these feelings and sensations are information.** For billions of years they, not symbols and images, have coordinated the global life system's growth, within which humanity evolved. If your upbringing deprived you of Nature's physical and mental joys, you may read about what you missed in **Earth Wisdom** by Delores La Chapelle. My experience shows that it's never too late to start. That's why many adults choose to participate in our outdoor expedition learning programs.

Electrical brain wave studies show that subconsciously we sense T-R feelings before each sentence we speak or each thought we express. What we **say** is a conscious rationalization or symbolization of how we **feel**. I **feel** disturbed by our prejudice against Nature, feelings and "lefties." That's why I write about it.

It is usually the natural world's T-R feeling-sensation part of us that is in charge of our thoughts and acts, since 85 percent of our brain is designed for experiencing, not for reasoning. Shut your eyes and think how long it takes for you to envision a pizza in comparison to the time it takes to add up the numbers $5+2+8+6+9$.

T-R is a law of Nature, an essence of how the Planet survives as a living organism. It is why, like modern fashions, Nature always changes, shakes and demands attention.

DRAWING CONCLUSIONS
We outline the Planet as an amoeba-like wavy circle to indicate that it is alive, immediate and consists of a fluctuating tension-building and tension-releasing (relaxation) pulse.

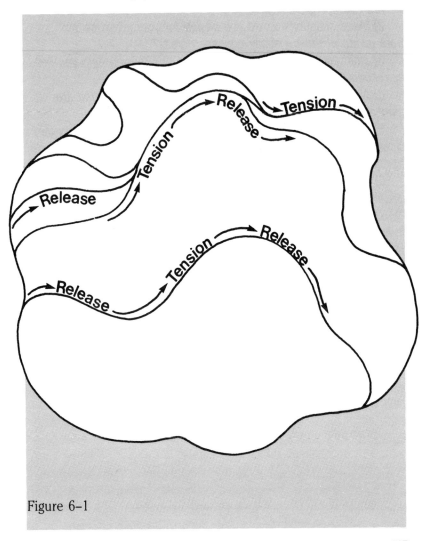

Figure 6-1

STUDY GUIDE—CHAPTER 6
By not taking the air for granted, the writer discovers new aspects of life.

1) Name five areas of your immediate environment that, like the air, you take for granted.

2) What situations would you design for yourself to become more aware of these visually unnoticed areas?

3) Hold your breath for a while. Does this tension help you better recognize your relationship with the atmosphere?

4) Name five other natural tensions you may sense that are connected to the Planet.

5) List ten different animate or inanimate objects. When do they experience tension and tension-release relationships? (Keep in mind that solid objects teem with atomic and subatomic motion.)

6) State five different ways that modern life separates you from sensing the natural environment directly.

7) Try to locate the feeling(s) behind each sentence you speak or each of your thoughts.

8) List some basic natural feelings and sensations that you have experienced on this day. They should feel less intense but no different than on the day of your birth. They are the voice or callings of Planet Earth through your Internal Nature. To what specific entities in Nature do they relate?

9) Close your eyes and sense the Earth's pulse in yourself. Imagine that your surroundings are part of that pulse, that you're actually sensing them until you open your eyes. At the moment you open your eyes, your surroundings change into what you see and what you have learned about the environment.

NOTES
On this page or elsewhere, write down any thoughts or feelings that have come to you from reading this and previous chapters.

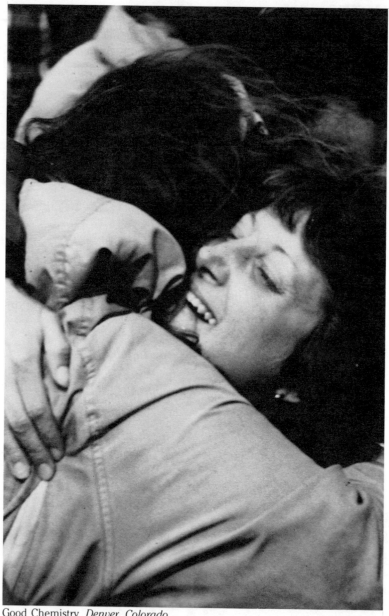
Good Chemistry, *Denver, Colorado*

CHAPTER SEVEN
THE AFFINITY BOND

A BBC television team recently convinced an Indian fakir to perform for them. The event promised to be a TV first. Before BBC's camera crew and a small crowd, the fakir played his flute above a large urn from which rose a rope, followed by a small boy who climbed the rope and disappeared into the clouds, pulling the rope up behind him. Ecstatic, the TV crew celebrated their "scoop." Upon developing the film, however, they discovered nothing on it but the fakir and his flute—no rope, no small boy climbing it. The fakir was accused of fraud.

The fakir replied that the crew, not he, was questionable, for they knew reality only by believing a mechanical black box rather than their personal experiences. "Wasn't it fraudulent for responsible media people not to trust and report what they actually saw?" he asked. He intimated that cultures and Nature were more harmonious before written words and black boxes replaced people's trust in their personal experiences and in each other.

Because our upbringing ingrains us with a black box viewpoint, we view the Planet mechanically. It appears to work like a machine. The discovery that the Planet regulates its own environment to stay within parameters allowing life to exist doesn't change the picture. After all, machines self-regulate, too. A thermostat turns the heat on when it's too cold and off when it's too hot. A toilet does the same thing with respect to the water level in its tank. We don't believe a toilet is alive, so why should the Earth be any different?

The difference, of course, between Planet Earth and a toilet lies not in their homeostatic mechanisms, **but in the source of these mechanisms.** Toilets are artifacts made **by people. We design their regulatory mechanisms** to imitate our own and Nature's ways. But Planet Earth's regulation is **self-organized.** Of its own accord, it functions as a life system. **In doing so, it needs powers of consciousness, communication and sensation.**

We would think a toilet were alive if we saw piles of clay and iron ore lying on the ground and observed them forming and firing themselves into porcelain and steel, shaping themselves into a toilet and its parts, fixing themselves when they broke and flushing themselves at the appropriate time. This would require consciousness and sentience and would be celebrated far and wide. Yet even the smallest seed of a common weed accomplishes all this and more.

People's original, principal adaptation for survival is their ability to make sense of the Earth's many different signals. These communicate the way the world works and how, like other species, we can relate to it for our mutual benefit. Learning in multidimensional ways incorporates the Earth's signals, the fakir's acts and the black box's recordings. Learning to know life from only the black box makes the fakir seem magical and fraudulent; it's like blaming beds for sex.

I learn about Nature from Nature. For we must allow ourselves to experience Nature and life's fullness in order to know them. Alone, neither the black box, words, nor any other single sense can completely convey Nature or life. And because definitions are only words, no definition of life exists or ever can exist. Therefore, since neither we nor the black box can define life or Nature, we really don't know what life is.

However, we are alive. We personify life. As living beings, **we must know what life is because we are it; we live!** But we have been intensively trained to believe that we can only know something when we "own" it in words and images, if we can communicate, reproduce or control it. This approach often denies the value of our feelings, our ancient life connections with Nature.

Because our learned desire for objectivity precludes the obviousness of our personal lives, we seek the knowledge of life outside of us in the laboratory or wilderness. Yet, neither alone is the best place to learn about Nature. For the closest natural area to us is not the nearest sanctuary, park or nature trail. **The closest wilderness to a person lies within himself;** our personal biology and the Planet's are one. It's like having your cake and eating it, too.

I sit on a stump listening to the Vermont forest. Occasionally a subtle breeze whispers in the leaves high above. The thrush sings, a red squirrel chuckles, the chickadee dee dee dees. Through the moist, sweet air, the whoosh from the passing raven's wings gently brushes my ears. But behind these woodland voices roars the overpowering stillness of the silent oaks, hemlocks and maples, the quiet glacial boulders, rock faces and soil, the hushed mountains, lake and sky. A deep quivering, awesome, hollow, echoing silence of sensation shouts throughout the landscape. It is the sound of all things in harmony, the song of Gaia, the Earth organism.

Sanctuary Much, *Yosemite National Park, California*

How sad that my language has trained me to take Planet Earth's silent concerto for granted, trained me to value instead the intermittent cries of wildlife and wind, to ignore the singing Earth's background of stillness. Authored by the ages, yet usually unnoticed, the unheard ballad of the valley's tranquility magically energizes me. The land's majestic quietude is the voice of a life force other than humanity. Like the virtue of the stars, its power escapes words.

One must participate in this sentient act called wilderness in order to know its peace. Where else can you find a Bible that instantly conveys the wisdom of five billion pulsating years of life? Where else can be heard the soft symphony of millions of organisms wisely expressing their mutual interdependency? This deafening silence and my pulse is that masterpiece.

No doubt exists concerning Nature's stress-healing power. Surrounding me during the past decades, it has taught me to sing with it, to revere the quiet melody that my upbringing labeled "vacuum."

Hear Affinity and Harmonize, *Joshua Tree, California*

In latter years, it has showed me that although seldom appreciated, the Planet music plays not only in the wildlands but in the Nature within ourselves as well.

Born into the cacophony of Western culture and dancing to its tune, we learn to reject the great silence within us. In the process, we lose its power. We label it boredom or loneliness and seek other amusements to replace it. In time our inner harmonic vibrancy disappears. Desperately we seek alternative stimulants to relieve the stress of our wilderness-displaced lifestyles.

We must combat destructive stress. Prevented from living outdoors by our social entrapment, we desperately need to gain Nature's powers elsewhere. Many express surprise to find that these energies are available from the biological wilderness within us, energies hidden in mental valleys by mountain-high words labeling them unimportant, boring, subjective, emotional, exceptions to the rule, unscientific, personal, insignificant, bad or wrong. By such labels, institutions addict us to their special interests rather than to the stabilizing sensations of the Planet lying in hidden valleys of our consciousness. The labels rob us of our natural survival relationships, the life desire of time.

I attempt to learn about the Planet from the wilderness within me. As I sit on this stump listening to the Vermont forest, I attempt to match my feelings with it. Tensions like hunger, thirst, suffocation, temperature, all have their satisfying tension-releasing counterparts in surrounding lakes, hills and sky. My relationship with them is that I exist because they exist. I deeply feel this.

After a while another feeling surfaces; I like to relate. I desire relationships and without them I would die. My survival depends upon supportive interactions such as relationships with parents, friends, employers, materials and natural systems. My desire to relate is not a weakness, a lack of self-sufficiency or independence. My desire for community, for a place where people know my name, is as much a biological urge as is thirst or hunger. Companionship is not extraneous but a necessity constantly needing fulfillment in some meaningful way.

Since the Planet and I are one, our congruency suggests that if biologically I desire relationships, every other wild entity in some fashion also desires them. People are part of the global ecosystem. To maintain relationships in a system, however, you must have some kind of communication with it. Otherwise, separately, the system does its thing and you do yours. Therefore, **Earth must somehow signal us and vice versa**. But how does this happen? We don't telephone or exchange newsletters. How do we communicate? How do we relate? How? How . . . I sit on the stump in the middle of a Vermont forest waiting for the answer I know is there—an answer I know is there because I have looked for it, because I have lived it, because I have felt it . . . And then suddenly it is there. So simple, so straightforward, so obvious, so beautiful. Just what you would expect from Nature. If I feel a positive-tension desire to relate to the natural world, that's a valid desire to relate. And since every other entity is similarly part of the system, they, in their own way, must also desire to relate.

It is all so clear. Each and every entity desires to relate, has an "affinity" for other entities. Affinity produces matter and stability. Oxygen has an affinity for hydrogen, water for sodium, fish for water and I for life. Two people or two minerals can have a "chemistry" between them that leads to bonding. Over time, the concept expands into four parts and still feels right:

1) All entities have an affinity for relationships, which by giving them more stability allows them a stonger existence.
2) Each new relationship forms a new entity, which in turn has an affinity for further stabilizing relationships.
3) The end result of the Earth's five billion years of affinity relationships is a pulsating, stable, self-organized global organism.
4) The global organism's affinity interactions organize, sustain and regenerate life.

Those four simple statements describe the way I find that Nature works to make possible life as we know it. It is what underlies the self-organized growth of plants, animals and geology. It is how and why, **without our direction**, affinity interactions heal a scrape or burn on our body or on the landscape, an embryo knows how to grow into an adult, and rust invades the scratches on my car.

Our biological nature is part of Nature in general. Nature in general seeks affinity relationships. So does our Inner Nature. Affinity is Nature and people's timeless survival drive. The Bible recognizes it: "Love worketh no ill to his neighbor, therefore love is the fulfilling of the law." (Romans 13:10)

How different this is from the scientific outlooks that reject the validity of feelings. I am reminded of Roger Sprung who claimed his dog didn't eat meat. "How is that possible?" I asked.

"I don't feed him none," he said.

At the IS THE EARTH A LIVING ORGANISM? Conference, Nobel prize winner George Wald repeated a conversation he had had with Albert Einstein in 1952.

> We were walking down the street when he turned to me suddenly and said, 'I have wondered for a long time how the electron came out to be negative . . . I have thought about that a long time, and all I could think was, it won in the fight!'
>
> The fight that Einstein was talking about was the fight between matter and anti-matter. You see, the rule is that when a particle of matter comes in contact with a particle of anti-matter, they mutually annihilate each other, and both are turned instantaneously into radiation.
>
> The idea is that though matter and anti-matter are perfectly symmetrical, though virtually equal numbers of each went into the Big Bang to bear this universe, when that enormous firestorm of mutual annihilation had finished turning that matter into radiation, one part per billion remained [that part won the fight], and that's our universe of matter. That's all the matter in this universe: all the stars, the galaxies and us.[6]

Could it be that what Einstein was saying is that one out of every billion "pro-matter" particles **desired** to remain or survive as matter and **fought** for its right to survive? Yes! Of course! Each of those particles' affinity for **being** is the same affinity that I feel for food, air, water and companionship. This I deduce by recognizing that we and the natural world are one! Independent feelings generated in a Vermont wood and in the genius of Albert Einstein mutually agree and converge across the years in observations of life.

The very act of **being** profoundly describes life's essence on both a particle and global level. "To be or not to be, that is the question," is the way William Shakespeare put it. But being is only accomplished through stabilizing relationships that support it. Being consists of an affinity or desire for such relationships. We must not take that desire for granted. Recognize who you are by saying, "I am a desire for water, air, food, love, warmth, beauty, freedom, sensations, life, community, place and spirit in the natural world." These pulsating feelings are Planet Earth, alive and well within you.

As we will discuss later, our upbringing buries the natural survival feelings found in each of us as it teaches us to reach for the material things in life. The affinity concept, however, suggests that **the essence of the Earth is not a material, it's an emotion,** a global communion, a fighting desire for relationships that survive. Thus gravity and magnetism should be considered emotions, even though our materialistic, objective perceptual habits can't see them as such. Yet we can sense magnetic personalities and grave situations. Although this concept is controversial, so is acupuncture. And if you're someone who questions acupuncture, let me ask you something, when was the last time you saw a sick porcupine?

The deteriorating state of the Planet, our environment and our lives clearly indicates that modern society has lost touch with the natural world's harmonies. Although Nature inside and outside us is identical, Nature is separated from us, lost in modern culture's complexity, just as a hiking student can get separated from our outdoor group and lost on a confusing wilderness trail. When trekking, we prepare for this contingency by agreeing that as soon as a person is missing, the group and the lost person will return to the last place we've been together. Most often this reunites us.

I have found that this hiking rule works just as well for reuniting myself with Nature. Nature and I meet again when I backtrack to rejoin Planet Earth's T-R affinities and then let myself feel and honor them. I then accurately symbolize the facts of my togetherness with Earth by realistically calling my feelings *Nature* and *using the two words interchangeably.* As I label my feelings "Nature," "Inner Nature" or "Earth," my level of comfort or discomfort at any given moment connects me to Nature and Earth's affinities. For example, I allow myself to say that when I feel hungry, Nature within me is signaling that it needs food; when I'm uncomfortable, my Nature is sending signals that it's being abandoned; when I'm comfortable, my Nature is signaling that is being supported. Similarly, I label each of my other mad, glad, sad or scared feelings as comforting or discomforting expressions of my Inner Nature. Whether I feel comfortable or uncomfortable, is Earth within me responding to the situations that touch it.

I've found that Nature's tension-building and tension-relaxation (T-R) dance within us communicates itself as varying levels and intensities of our Nature's state of well-being. We can diagram this comfort-discomfort phenonmenon as follows, using our affinity for water as an example.

Affinity:	Affinity for water	
	▼	▼
Affinity Status:	consummated	thwarted
	(water available)	(water unavailable)
	▼	▼
Your Nature Expresses:	comfort	discomfort
	(relaxation)	(tension)
	▼	▼
Your Expressed Feelings:	I'm fine, O.K., not thirsty, content, comfortable	I'm thirsty, not O.K., anxious, scared, concerned, tense, unhappy, etc.

The diagram holds true for each of our other Earth affinites such as that to food, warmth, air, mobility, love, territory, companionship, sensation, etc.

When you know your feelings and call them Nature, you empower your cognitive self to recognize Nature. This connects your rationality to the natural affinities within and around you. A new togetherness blossoms when these long-separated parts of you fuse. You know that not only Earth is a sentient global being, but that each of us embodies its affinities and always senses them on some level. When you feel the Earth, you're reunited as kin because you have the same survival interests.

You can easily gain Earth's support for yourself each day by writing down your feelings and labeling them as your Nature expressing its comfort or discomfort. Try it. Write down how you feel right now and also your feelings about the different people, situations, readings and activities you experienced today. Write down your observations of other people's real or televised situations and how they felt in them. Then attempt to express each comforting or discomforting feeling or sensation you have noted as an expression or function of Nature. For example, fill in the blanks with *one* of the words in the parentheses:

My (or their) Nature was _____ (supported, attacked, threatened, acknowledged, frightened, accepted, etc.) and it felt _____ (uncomfortable, comfortable).

My Nature desired _____ (food, water, shelter, relationships, habitat, mobility, etc.) and I felt its discomfort until this affinity was consummated.

Their Nature was _____ (enhanced, abandoned, nurtured, stressed, demeaned) and it felt _____ (comfortable, uncomfortable).

When the boss _____ (criticized, praised, hired, fired, promoted, etc.) me, my Inner Planet experienced _____ (abandonment, support) and I felt _____ (uncomfortable, comfortable).

You will find that, in time, your Nature will express gratifying confidence in you just because you honor its affinities and guide your life accordingly. Connecting your affinities to the timeless pulse of Earth increases your stability. And you'll find that when friends notice this new connection in you, speaking with them about it entices them to develop their own Earth kinship. But you verbally have to use Nature and feelings interchangeably to make it happen; the idea alone isn't enough.

As I watch animals and plants go about their daily lives, I recognize that they react to stimuli that I can't sense. My dog hears, smells and knows his surroundings in baffling ways. The Earth organisms's flesh and blood swarms with interconnecting affinity stimuli and sensations that have evolved over Gaia's multibillion year life. That is part of each entity's "spirit." All living things plug into some of these stimuli and react accordingly. My professors knew such actions existed, but they taught me to call them instincts. They also taught that each organism contains a chemical memory that "mechanically" directs each organism's life instincts.

My professors' "dead Earth" story is only part of the picture. The remainder of the story is that "instinctive" behaviors also rise from sensitivity to the multitude of Gaia's affinity stimuli that pervade the environment. That communion interconnects life globally.

The conclusion of this tale is that life's evolution or direction may be based upon the degree of environmental consciousness that organisms or species possess. The greater an organism's conscious affinity for its environment's and its own workings, the better its ability to integrate itself successfully into Gaia's life. The fittest may be the organisms with the most developed ability for global and local awareness, for greater awareness allows for greater affinity flexibility. That quality, interestingly, appears to increase from the "lower" to the "higher" plant and animal species. Increased consciousness is necessary to establish affinity relationships, as more diverse organisms emit more diverse and complicated signals. The process is not planned. It's effected by immediate relationships and ongoing self-organization. It is a massive "pulling yourself up by the bootstraps" process, a global community effort to adapt to the sun's whims.

Cosmologically, the heart of the matter is that no difference exists between what we call living and dead entities. Each form of matter and wave is an expression of a universal affinity that forms stabilizing relationships. When observed apart from the whole, some entities interact with motions and sensations similar to those we humans experience. These entities we call alive. Others react so slowly that, unlike ourselves, they appear motionless and without sensation. These entities we call dead. Thus "dead" sea water is almost identical to "living" plasma in our bloodstream. And a "living" giant redwood tree is 99 percent "dead" wood. But beyond our arbitrary, self-centered standards, everything appears to act as the flesh and blood of an affinity-based cosmos. It's as though a force can strike a red object and be expressed as red (dead) particles, or it can strike a green object and be expressed as green (alive) particles. But without the original force, we would recognize neither red nor green particles. Affinity's expression is the force that produces the universe. All matter experiences it on some level. One way we experience it is through the survival feelings we call "love."

Consensus is an essence of Nature's affinity network. In the long run, unlike modern technologies, Nature works very slowly, giving every entity time to exhibit its nature, to consummate its affinity partnerships or make new ones. In that way, Nature harmonizes. It produces no garbage because all individuals in the natural community, from atom to continent, are valid and essential, even mosquitoes, leeches and poison ivy.

* * * * *

How can a global life system be founded on affinity and end up with competition amongst its members? It can't! Competition usually exists if the viewer's consciousness perceives the Earth as inert resources to be exploited and ignores the survival of the whole Earth as a living organism. Like other social ills, competition is a perceptual side effect of not recognizing that Earth is a living organism grown and nurtured over the aeons by all entities consummating their affinity desires.

Competition underlies most of our relationships even though we know that it is a questionable motivating force. Competitive feelings overwhelm the more subtle values found in life relationships and activities. Little League baseball is like World War II with innings.

Competition is a long-accepted and documented view of Nature. From Darwin to any modern ecologist, competition is the solidly based explanation of why some plants, animals or people are able to survive in a specific habitat. The key to survival, evolution and ecosystems is that the best competitor for water, nutrients, sun or air is the one that survives. Why shouldn't competition be an accepted part of human behavior? People are natural animals. People feel competition. It is part of us. Why deny it or disapprove of it?

However, if we experience the process of life as being based on competition, we encourage competition. The result: people relate to other people, places and life forms, competitively. Competition leads to exploitation, isolation and strife, the antithesis of conservation and peace. Because of these negative effects, competition should be rejected as a reasonable way for people to relate to the Planet and to each other.

Consider competition as it exists in Nature. In Nature, we find that competition between organisms, and between species, stems from two basic relationships that are counterparts of each other:

1) Each organism and species produces many more offspring than are needed to replace themselves; these progeny compete for habitats, niches, nutrients and survival.

2) The Planet has a limited amount of natural resources, and all organisms must compete for these resources in order to survive.

Such observations lead the naïve observer to conclude that the removal of some organisms from a population will place less competitive pressure upon those that remain. Life would, therefore, tend to flourish and be stronger for their removal. Yet, actual observations prove the opposite to be true. In the long run, the removal of species from the ecosystem breaks the web of life and weakens the system.

We must build theoretical models of Nature that meet the observed facts. In one model, when we arbitrarily eliminate use of the concept of competition, not only does the model work, but it is also congruent with our observations and experiences in human communities.

The key factor in the design of an acceptable model is that we replace competition with its affinity antithesis, cooperation. Our reasoning is based on three lines of thought which we are forced to use because we have excluded competition:

1) The overproduction of offspring in living populations produces **stress** on each individual and species.

2) The organisms or species that best **cope with stress** are the organisms that survive.

3) The organisms or species that best establish **cooperative affinity relationships** with each other are most adaptable to stress and are, therefore, most likely to succeed in any given habitat or life zone.

I must emphasize that fitness and adaptability are **based upon the ability to cooperate,** not compete, with the global life system.

Although this differs vastly from Darwin's competitive "survival of the fittest," we must recognize that much of Darwin's research was done on islands. Islands are easy to study because they are isolated from the mainland. But this isolation also makes them atypical of life in large, continental populations which are subject to the tension, effects and guidance of many more continental microorganisms and cumulative forces than are found in an island's isolated, limited population. Thus some of Darwin's island findings are atypical of global life's workings.

Sometimes the best way to cooperate is for entities to disband ongoing affinity relationships and instead make new ones that are more globally fulfilling and life-supportive in the long run. We often call that process "death," instead of calling it "changing from one life system to another."

Competition stems from devaluing affinity feelings. This leads to anxiety about the availability of raw materials or money for overprotection from Nature.

These very same observations were made by the Shoshone hunter-gatherers. According to Peter Farb, they did not compete or war over territory because **"territory is valuable only at those times when it is producing food, and those were precisely the times when the Shoshone cooperated for harvesting, rather than made war."**[9]

A society is a means of coping with the normal stress of survival in Nature. In time, a society can meet its survival goals through interpersonal and environmental cooperation. The Amish, the Hopi, some organic farming communities, our expedition groups and populations of social animals such as wolves, exemplify cooperative societies.

Seek the Sacred Solution, *Bear Butte, South Dakota*

A museum educator in Bear Butte, South Dakota, explained the dynamics of cooperative affinity relationships. He told us that one hundred years ago, when the immense buffalo herds spanned the plains, there was more sustained yield of meat on the hoof than there is today, despite our modern practice of cattle farming. There was also soil build-up rather than erosion.

The curator said that if today we were to bring back the buffalo herds and wildlife that were once here, we would find we could not keep them alive on the plains because we have destroyed the prairies, the incredibly nourishing soil and soil-building plant communities that were needed to sustain the herds. "First, we would have to bring back the proper plants," the curator explained, "but we can't do that because those plants are dependent upon relationships with large herds of buffalo for aeration and fertilization." Today the land is over-cultivated, overgrazed and eroded. We have replaced the ancient buffalo with modern bull.

We asked if there was no solution to this circular problem. "Not a simple one, no," he replied. "If a solution were to be found it would have to be cultural. It would have to establish cooperative relationships between people and Nature. These would allow Nature to take its complete, slow-but-sure course of building up the soil, the buffalo and human populations as a unified whole." That solution would develop wisely civilized cooperative relationships, and not let the Earth continue to be a high-risk neighborhood. What is crucial is that acts, not mental manipulations alone, take place that enhance global life's existence. Unless people are prepared to make this commitment, however, we will not have a home where the buffalo roams.

At no cost to us, the Planet's cooperative affinity relationships maintain life and clean up our biodegradable wastes, too. The cost and size of an Earth-governing computer that we might otherwise design for this purpose would be outrageous and its technology impossible. But why bother? The Planet itself is a self-made computer. Each of its entities contains a portion of the knowledge of how the world works. The universe's stabilizing affinity program for belonging coordinates them.

There are, unfortunately, distinct differences between the way our Inner Nature actually works and modern society's instructions on the subject. These differences cause our personal and global stress, a subject we shall explore in Part Two. Meanwhile, keep validating your comfort and discomfort in order to reunite with Nature. For example, with respect to this chapter, ask yourself, are you more comfortable with: a competitive or a cooperative world? a world harmonious with, or exploitive of, Nature and yourself? the Planet's essence being an affinity emotion or an uncaring molecular accident?

DRAWING CONCLUSIONS

Every entity, be it subatomic particle, planet or star, (a), has or is an affinity for stabilizing T-R relationships with other entities. We depict these affinities with blue arrows between the entities (stars) (b). Once a relationship forms, (c), this new relationship has an affinity to relate. We depict this with arrows from the relationship arrows (d). Over time, pulsating T-R affinities form the Planet.

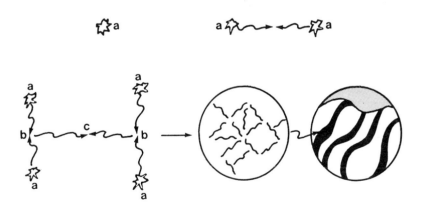

Figure 7–1

STUDY GUIDE—CHAPTER 7
In this chapter, self-regulation (homeostasis) is recognized as a means to identify life.

   1) Identify ten places in Nature and in your personal biology where non-artificial relationships organize and self-regulate.

   2) Read some accounts of how children were brought up by wild animals and survived.

   3) How do you think you know something is alive, if life can't be defined?

   4) Have you ever sensed companionship when alone in a natural setting? What might you have to do to "revere the vacuum within yourself?"

   5) Identify your personal affinity relationships. Distinguish those which you believe are natural.

   6) List situations where you feel competition is destructive. Could these situations become cooperative? How?

   7) Henry David Thoreau said, "Wilderness [Nature] is a civilization other than our own." Permit yourself to find or imagine a plant, an animal and a rock that you like. In turn, let each of them talk to you, explaining their affinities, how they and their civilization operate. Take their words and apply them to yourself and your society. What similarities or differences do you find?

   8) Consider stress in your own life and how you deal with it. Does it make you comfortable or uncomfortable? Do you compete or cooperate? Are you conscious and concerned about survival or about personal possessions and prestige instead? Locate some personally stressful relationships and map out how cooperation with Nature or with people might reduce your stress and increase your survival.

   9) List six forms of garbage or waste created by your daily life. How do you feel about them and their disposal? In turn, imagine yourself to be a plant, animal, mineral or natural event. What are your waste products? How are they disposed of? How might you feel about them as food for your neighbors?

10) Choose ten natural objects, situations or places that give you pleasure. Explain why they are pleasurable. Add your explanation to the sentence, "I am a person who gets pleasure from _____."

Permit yourself to thus recognize yourself.

11) Observe how affinity can produce unity. Stand in a circle of twenty or more people. Have each person select an active motion, like waving their hand or shaking their head. Also have them pick a person across the circle from them and imitate that person's actions (affinity). On a signal, have everybody start doing their motion and then imitating. Does the whole group end up doing the same thing? Sometimes!

12) In the chapters that preceded and follow, study your notes at the end of the chapter. Identify whether they make you comfortable or uncomfortable and locate how or why your Inner Nature is giving you these feelings. Validate these feelings for, unless they're misled, they have meaning.

NOTES

On this page or elsewhere, write down any thoughts or feelings that have come to you from reading this and previous chapters.

The Art of Survival, *Glen Canyon, Arizona*

CHAPTER EIGHT

THE SECRET LIFE OF PLANET EARTH

Once I seriously recognized the erotic universality of affinity's Tension-Release process of relating [T-R], and the meaning of my comfort-discomfort levels, I underwent a dramatic change of perspective with the world around me. Suddenly, Planet Earth and I, like Siamese twins, shared a common backbone, physiology, metabolism and consciousness. Identifying with the Earth did not make me unrecognizable as a person; rather, it strengthened my self-concept.

I could see that at the moment of birth, we have no modern symbols, images or words. Yet we have two mothers: our human mother and our planet mother, Earth. The infant is the Planet personified; it embodies the Planet. It knows pulsating survival affinity sensations and feelings, the same as Earth. They exist to acquaint the infant with the natural world's life-supporting elements, the Earth Mother. It is only later that our human mother and society teach us the other part we need for modern survival.

The infant senses the affinity tension called thirst because the Planet has rain, lakes and rivers to give it water. The infant's thirst tension is released when it drinks water and is not satisfied by a glass of dust. Thirst is an affinity feeling embracing the Planet's water, and, in turn, the infant's urine satisfies the Planet's affinity tension for liquid and nitrogen.

The infant senses the tension of suffocating because the Planet can provide it with air. Breathing releases this tension and satisfies the Planet's tension for carbon dioxide.

The infant senses hunger because the Planet wants it to eat and live. The infant senses the tension of excretion because the Planet needs its by-products as food for other organisms.

The infant desires mobility to move toward or away from environments that can help or harm it.

The infant feels loneliness because it craves the life-supportive niche that the Planet provides. The infant satisfies the Planet's desire to enter stabilizing relationships.

The infant senses temperature because it seeks environments that best support its life, and vice versa.

The infant experiences the affinity tensions of sexual desire as part of Nature's love for it and its kind.

The infant senses music, form and color so that it may react to those aspects of the Planet. The infant satisfies the Planet's desire to be appreciated in a relationship.

The infant loves life because love is the nature of life. The infant may experience many other Planetary sensations, too, as noted in Chapter 5.

Trust Survival Feelings,
White Mountains, New Hampshire

Spirit of the Inner Child,
Canadian Rockies, British Columbia

The above are each individual's natural Planet affinities. Throughout life they remain alive in each person because they are life. All our other feelings, creativity and desires evolve from them.

Interestingly enough, at birth and thereafter, the infant, when allowed, meets its interests as well as many adults can. If, as sometimes happens, a wolf, gazelle or other wild animal raises the infant, it survives by developing some of its wild parent's adaptive characteristics and behavior. A child mostly learns to know Nature from Nature, not from a classroom or book. Over the aeons, the human brain has evolved to register experiences and emotions, not to learn only from abstract symbols and rational thought. To be fully rational, our rationality must honor this fact.

All human feelings and endeavors derive from Planet Earth's T-R relationships. Natural human feelings are expressions of the Planet's T-R in people. You can always trust positive signals and feelings about Nature, for they always contain survival value. In concert they create the love of living, simply desiring to survive because life feels good. On some level, whether we can see it or not, all life forms have some kind of T-R survival signals.

The infant within us never disappears. Most of us don't recognize that it is our Inner Nature and is entirely sentient. It functions by relating comfortably or uncomfortably to the global life system's T-R.

Because our Inner Nature Infant experiences only sensations and feelings, it doesn't differentiate between feelings that have been derived from seeing a real object and those derived from a symbolization (word or image) of that object. Feelingfully, our Inner Nature knows seeing a real tree and the word "tree" identically and reacts accordingly. For example, the words and images of an adventure movie evoke adventure feelings in us, and a song about a long lost love can evoke the same emotions as the real thing.

Our Inner Nature relates to modern society's advertising, artifacts and built environments as if they were parts of a wilderness area. It senses them and often binds with them in a love/hate relationship. It seeks them for their survival value, for the mutually beneficial wilderness T-R affinity relationships that ordinarily maintain life. Our

Inner Nature is then manipulated by them to meet society's goals. Unbeknown to us, during the time we live out our daily lives in modern society's artificial world, our Inner Nature is constantly experiencing supportive or threatening wilderness relationships there.

Many lifelong uncomfortable, bad feelings are first known through the modern birthing trauma. We relive these feelings when new threatening situations trigger them. It almost seems sometimes as if God gave us our Inner Nature so that the advertising industry could disturb it, and then sell us products to make it feel better.

Our comfort-discomfort feelings at any given moment entirely depend upon what symbols, images or situations our Inner Nature is reacting to at that moment. We either heed our danger feelings or we pacify them with images accurately or falsely portraying safety. It's like the heartbroken child, who, while visiting many department stores at Christmas, has discovered a Santa Claus in each of them. "Santa isn't real," she sobs. "Yes he is," replies her father, who turning to the nearest Santa says, "Santa, tell her about the time you fell into the Xerox machine."

* * * * *

As red morning sunshine brilliantly paints the clouds and mountains, so each day, dawns a new light on humanity's relationships with Planet Earth. Presently illuminated is the recognition of Planet Earth as a global-sized sentient living organism, whose T-R affinity successes and failures we embody and experience as comforts or discomforts.

I find that what we call imagination, consciousness, sensations and feelings are vital survival mechanisms of the Earth as well as of people. Over the years, outdoors, they have introduced the Planet's secret life to me. Similarly, I introduce it to you here, the coursing life only thought about and hinted at in the previously discussed scientific reasoning known to us as the Gaia Hypothesis. This is the life the scientists are looking to find. Here it is, ready to become a part of you.

Take a few deep breaths to consciously reacquaint yourself with the Planet, in its language. Note that inspiration and respiration are functions of Mother Earth's ancient wind-spirit, that part of the Planet we call "atmosphere." Savor a few deep breaths and feel the spirit.

1) Spend as much meditative time as you need with each of the following statements and paragraphs. Note that the pause following them, indicated by three dots . . . is as important as their message.

2) Enjoy trying to make some space for their sentient meaning. Note whether they feel comfortable or uncomfortable.

3) Make this a game if it feels foreign. You don't have to agree with each paragraph; just try to understand it fully and be comfortable with that. Tax your imagination and creativity. Then relax (T-R). Then continue.

Ready?

Imagine an apple: its color, taste, smell, texture and sound . . .

Imagine yourself getting smaller and smaller until you can enter the apple through its stem . . .

Work your way into the center of the apple . . .

Once there, slowly allow yourself to expand until you occupy every part of the apple . . .

Think of yourself as part of the affinities, feelings and essence of the apple. Become part of its nature . . .

Remain as one with the apple and slowly let the apple enlarge to basketball size . . .

Enlarge with it and keep enlarging it to room size, house size, town, then county size . . .

Slowly you and the apple become bigger until you're larger than the state, the country, the continents and oceans . . .

You become the size of the Planet. Try to feel comfortable being the size of the Planet . . .

And now that you are the size of the Planet, merge with the Planet. Slowly let yourself flow into the life-dance of Planet Earth on its orbit around the sun. Become the Planet as you did the apple . . . [If this imagery does not work well for you, stop here and take some time to develop thoughts and feelings that make you at one with Planet Earth.]

Now give yourself permission to learn from your human life about the Planet's life and vice versa . . .

Experience yourself as a feeling, a desire to relate that is rewarded by each relationship's gratification . . .

Let your hands touch each other, and then your face. By affini-

Flesh and Blood in Common, *Mt. Robson, British Columbia*

ty sensations like touching and feeling, tasting, smelling, hunger, thirst and hearing, not by words, does the Planet know itself and survive . . .

Know yourself as a symphony consisting of infinite numbers of affinities interacting and building over the aeons . . .

Press your hands on your temples and feel your heartbeat. That pulsating tension-relaxation essence of your life still beats as it has since your birth billions of years ago . . .

Desire the sun's presence. Inhale sunlight and create oxygen from it. Let yourself hear and feel your breathing. Your planetary breath circulates air through plants, animals, water and minerals, making their lives and your life possible . . .

Imagine the sun becoming warmer and warmer and you tensely sweating thunderstorms and hurricanes to cool off. Imagine fanning yourself with arctic air or cooling yourself by running an arctic ice-water bath . . .

Imagine yourself erecting an umbrella of clouds or a blanket of carbon dioxide, or sunshine reflectors of daisies, snow or glaciers that you can use to either reflect or hold the heat . . .

Feel hungry and satiate your hunger with sunshine, your major food source . . .

Think about your lips drying. Lick them with rain to make them feel comfortably moist . . .

Put your arms around yourself. Enjoy being embraced and fondled by the universe and yourself . . .

Feel relieved that you have organ systems satisfying your need to excrete. Compliment yourself for discovering how to recycle your excretions into healthy food and water that you safely eat and drink again. Be proud of that achievement . . .

Be in awe of your life-giving respiratory processes that occur slowly in decomposition, more quickly in mammal metabolism and like a flash in a forest fire . . .

Enjoy your sexuality, your planetary desire for life to continue. Enjoy being a fertilized, growing egg of the universe . . .

Imagine an overabundance of saliva in your mouth. Swallow it like the Planet swallows carbon, salt and methane to regulate the levels necessary for life's maintenance . . .

Imagine desiring harmony within yourself and knowing how to maintain it by altering yourself continuously . . .

Imagine yourself as the womb of life; feel the excitement as each new moment dawns . . .

Feel companionship, a sense of place, support and belonging, a sense of being whole and important for life's existence . . .

Celebrate being something very worthwhile. Rejoice that your survival desires validate all your natural processes; you are never bad, wrong, negative or evil . . .

Feel nurtured and nurturing. Feel musically harmonic like a simple folk song or a symphony . . .

Feel wonderful that you are conscious to enjoy all these aspects of your survival. Be happy you are them. Delight in their consciousness of you, their worship of your benevolence, leadership and wisdom . . .

Feel secure that even as you sleep, the life process maintains itself in celebration of you . . .

Dream of creating new organisms and life systems that share, support and enjoy your life, that will enrich life for others . . .

Congratulate yourself for brilliantly establishing life over the ages without using written words, numbers, clocks or money . . .

Enjoy your ability to heal yourself when you're injured or under cultivation's stress and tension . . .

Feel vital and active because you consist of ever-adjusting relationships . . .

Look out into the stars and sense the wonder of looking deep into yourself and your beginnings . . .

Enjoy being illiterate . . .

Thank God for having created your enchanted life and potential for enjoyment . . .

Let yourself experience your essence. Exhale and don't breathe. As you hold your breath and the discomforting tension builds, recognize that you are experiencing your survival voice, a language that knows neither words nor sentences. Feel secure knowing that beyond yourself is an affinity force revering your importance, that loves you, that insists you breathe even if you choose not to. Now inhale and release tension, reconnect your totality by breathing . . .

Recognize that you breathe both air and sunlight . . .

Feel powerfully confident, knowing that your universal affinity language of tension building and tension release communicates growth and survival for every single entity in your global life system . . .

Celebrate the delights you receive from music, sound, color, texture, tastes and design. The global community gave you these birthday gifts so that its life would be known to you . . .

Enjoy the pulsating signals between yourself and the moon, sun and universe. Enjoy the tickling of your tides, the streaming of your rocks, gases and fluids, your volcanic burps and your harmonic peaceful song of unity . . .

Feel honored that your survival is the basic origin and purpose of the thousands of differing human cultures . . .

Love life, because you are life . . .

Love being . . .

Feel unified as you experience all the matter and forces of yourself as manifestations of your basic desire to live . . .

Delight in knowing that the God some people put in Heaven is a reflection of your life; revere the fact that most people worship you as either their mother or God . . .

Appreciate that whatever happens to you as the Planet, also happens to you as a person and vice versa. Your acculturation aside, you and Earth are identical . . .

Recognize that this imagery feels familiar because the Planet's life and you are one. It doesn't work in becoming one with a machine, because a machine is neither naturally sentient nor conscious, nor does a machine organize, perpetuate or regenerate itself . . .

Be aware that as a living organism you maintain your life by relating to the surrounding solar system and universe. For your survival, you must cope with their fluctuations and forces. Sense yourself as a global organ community exactly like your human organ community. Counterparts of your kidneys, liver, pores, stomach and heart exist in your geography. Habitats function not only as homes for their plant and animal inhabitants but also to sustain your global life needs . . .

Your metabolism is driven by ingesting the sun's high energy radiation and excreting low energy into space . . .

Your ocean's kidney is the activity of the continents, oceans and corals. Their interactions form warm, shallow inland seas that evaporate water into the atmosphere and crystallize out excessive salts and sediments. These become part of your mantle. Glacial melt and water storage also modify your salinity . . .

Tropical rain forests and phytoplankton act as kidneys to cleanse your atmosphere, as they remove carbon dioxide from it . . .

The continents and atmosphere are your liver. They store minerals and gases that you use as food . . .

Your lungs are plants living in the land and sea. They breathe in sunlight and carbon dioxide. They exhale oxygen and water vapor . . .

Your circulatory system is the ocean's and atmosphere's currents, the rain and the mountain and valley water systems. They distribute your food and gases and help regulate your temperature . . .

Your digestive system is the erosion of rock by lichens, climatic changes and glacial action . . .

Your sweat glands are the deserts, volcanoes and clouds, as well as your color. They release your excess heat into the solar system . . .

Like plasma and lymph, the solar wind bathes you . . .

Your heart is your rotation on your axis causing the heart-pumping action of daytime and nighttime temperatures. It is also your angle to the sun and your orbit around it that cause seasons, polar ice caps and equatorial tropics. Heat differentials produce your circulation externally and internally . . .

Your musculature is your flowing inner core and continental movements . . .

Your skeletal system is your continents, mantle and inner core . . .

Your beaches act as fingernails, protecting your weaker coastal lands from wind and wave . . .

Your skin-healing scars and scabs are your hardened volcanic plugs and lava flows . . .

Your nervous system is your networked electrical, hormonal, gravitational and magnetic forces, as well as the networked T-R community sensation dynamics and waves that interconnect all aspects of yourself . . .

Your forests act as hair, insulating and protecting your body . . .

Your heat regulation is accomplished through limestone, oil and coal formation from bogs, tree roots and sea shells. These life forms bury carbon and thereby thin or thicken your insulating carbon dioxide atmospheric blanket. Storms and hurricanes redistribute your local excessive heat. Glaciers make the oceans retreat, providing new forest areas that consume and bury carbon. Glaciers also grind rocks

Planet Bones, *Cutler, Maine*

to dust which the wind spreads globally. Rock dust fertilizes your forests, thereby increasing their breakdown of rocks and carbon dioxide intake. Glacial movement and ocean sediment deposition give you backrubs that squeeze your insides, perhaps forming heat-releasing volcanoes . . .

Your water regulation is accomplished by water storage in glacial ice, snow, soil and clouds. You make new water in your mantle, injecting it into the seas at your rifts. Your mountains act as breasts, tapping the clouds for nurturing moisture which they store and distribute over time . . .

Your endocrine systems may be climate and weather patterns: humidity, glacial and volcanic dust and erosion . . .

Each and all of your creatures and entities are your sensory organs . . .

Your oceanic rifts help regulate your temperature and salinity as they alter the size and place of your heat-collecting oceans . . .

Your brain is the shared sensation network of the T-R affinity community and the God who created your desire to live . . .

Your skin and cell membrane is the five thousand mile layer of atmospheric and stratospheric gases, magnetic lines of force, radioactivity and electrical waves that cope with the solar wind and universe . . .

Your spirit is the desire to relate beneficially that lies in every entity of your being . . .

Feel secure knowing that you consist of the six billion years of knowledge and relationships of your entities. They know how to perpetuate and regenerate you. Rejoice that you are part of them and that they want and need you, that you personally relate to them by becoming a different entity or lifesystem upon your human death . . .

Delight in knowing that you are as wise and lovable as The Extraterrestrial. You are E.T. and you have his powers when you feel "at home" with your global community . . .

Nobody has seen you have offspring. Perhaps you reproduce by creating viruses and other life forms that could fertilize other planets when they are ready for it. Or perhaps you have not yet reached puberty. Although alive, maybe you are not yet mature enough to reproduce. Maybe your reproductive organs are our space programs . . .

People are an embodiment of you. You become more conscious of yourself through their symbols and images. You are doing that right now as you read . . .

The accuracy of your consciousness in conveying to you the true nature of your global life system's ways is a major adaptation for your survival. It allows you to continue to cooperatively grow with the life community as you have since its inception. Biologically, you are a hologram of the Planet's nature . . .

As a final image, fly out from the Earth, look back and think of the Planet as a womb of life as we know it. Look closely, and perhaps mixed in with its flowing placental clouds, continents and waters, you may see your adult self in the prenatal position. Today, you, I and the Earth still share the same heartbeat. And, as we touch the Planet, we touch each other . . .

We discover by becoming one with the Earth, that Nature within and without is identical. **You can portray this relationship by letting a pencil represent Nature. Place one end of it in your fist. There it represents Nature in you. Note that when you hit or move the section of the pencil protruding from your fist, you can feel the impact or motion within your fist. Notice, too, that when you press your fist's fingers against the pencil, the protruding portion is pushed around, as well.**

Like our effects upon the pencil representing Nature, when we stressfully disturb Nature outside, our Inner Nature feels the disturbance within us. Our Inner Nature becomes stressed. Likewise, when our upbringing manipulates our Inner Nature, we often react by stressfully manipulating the natural world that surrounds us. The corollary is also true, when we treat Nature within or without with respect and love.

* * * * *

This imagery best conveys in writing the Living Earth as I encounter it in Nature. Every nature trail and outdoor activity should embody it. If you find, as I do, that it makes you feel good, read it again whenever you so desire. As part of the Earth, in time you can learn to feel its vitality continuously.

Note the portions of this imagery that cause you some discomfort, and see if you can identify which of your Inner Planet's affinities these parts thwart. Is the threat real or imagined?

Treat the Earth as if it were dead, and it looks and acts dead. But treat it as if it lives, it acts alive. This may occur because by admitting the possibility of improbable events, we increase their probability of occurrence. For example, my attitude and actions can determine whether a wetland becomes a parking lot or remains a nurturing womb of life. With this in mind, we should definitely protect Planet Earth under the Endangered Species Act. As far as we know, it's the only living organism of its kind. Furthermore, its well-being unifies people with Nature and each other. (See Epilogue.)

Sometimes I think of the Planet as a cross between a tiger and a parrot; I don't know exactly what it is, but when it speaks, I listen.

Planet-Person Signature, *Mt. Rainier, Washington*

DRAWING CONCLUSIONS
The blue map of the Planet is identical to that of a person and is so labeled.

DIAGRAM OF PERSON AND PLANET

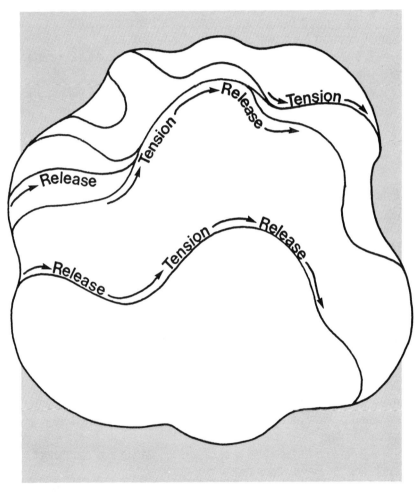

Figure 8–1

STUDY GUIDE—CHAPTER 8
The writer claims that people are born biologically as children of the Planet. The infant within you never disappears.

1) Do you recognize the infant who lives inside you? What things might you do to become more aware of it?

2) Try reading this guided imagery to a friend or friends. Sometimes you learn best by teaching somebody else.

3) Imagine E.T. to be a representation of the whole Living Earth and with that in mind, see the E.T. movie again.

4) Visit and compare natural areas and people's T-R nature. Try to identify what planetary life role they are fulfilling at any given time.

5) Try to sense how the Nature within you or others is feeling or responding within any given situation.

6) Try to recognize how any advertising attempts to disrupt your Inner Nature and have it respond by seeking the advertised product.

7) When you were a young child, much of the time you directly experienced your Internal Planet. Can you remember some of these feelings now? Try to sense how you felt as a child or feel in general during various times of the day. Joyful? Depressed? Anxious? Hungry? Your comfortable feelings signal that your Internal Planet believes it's receiving support; uncomfortable signals signify that you're threatening it. List the situations that elicit these feelings. Which aspects of the natural world do these feelings come from? Where might they be found in Nature? What is their survival value?

8) From womb to tomb we usually learn who we are from our parents' and other people's validation of us. Without validation, many of our important feelings about ourselves only exist as unrooted daydreams or wishes. Have a friend help you validate your global identity.

   a.) Sit face to face with a close friend, staring into each other's eyes, and for five minutes repeat the statement:
   "I am _____." Fill the blank with

whatever concepts you believe are representative of you (for example, I am a "concerned person," "musician," "habitat," etc.). Have your partner's only response be "thank you" to each statement you offer. In your statements, integrate concepts you feel comfortable with from this chapter. Then switch with your friend and become the "thank you" person.

b.) Upon completing this, reverse the exercise. For five minutes say to your partner, "You are _____," repeating to your partner what they told you about themselves and have them again say, "thank you." Then switch roles.

9) Go back to Chapter 5 now and attempt to find in it scientific data that supports the statements in this chapter.

NOTES

On this page or elsewhere, write down any thoughts or feelings that have come to you from reading this and previous chapters.

Civilized Behavior, *Beacon, New York*

PART TWO

THE CIVILIZATION OF NATURE

Cultural Filters Bias Perception, *Molas Pass, Colorado*

CHAPTER NINE
THE TROPICMAKERS

While in the Adirondak Mountains on her first camping trip, Madeline, a woman in our expedition group, offered to fill the canteens with river water. For safety, we told her to use water purification tablets, which we gave her. Surprisingly, when we drank the water, it didn't have the typical iodine taste that accompanies the tablet's use. "Did Madeline actually use them?" we wondered. She insisted that she had. "But why no iodine taste?"

"I don't know," said Madeline. "I threw the tablets in the river and then filled the canteens. Right?"

In May of 1986, I return to the Colorado River, a heart stream of America's southwest wilderness. I return there to purify myself, not the river. Our modern ways had already "purified" the river. We'd cleansed it of its wildness by building a massive power dam on it, flooding 2,500 miles of canyon shoreline in the bargain. With deep inner conflict, I visit what remains of Glen Canyon. Its monster generating plant at Page, Arizona chokes the life pulse of the Colorado, drowning God's finest sculpturing of the red rock landscape in Lake Powell.

I go to Glen Canyon to visit the yet unflooded upper portions of its relatively intact side canyons: Twilight, Cathedral, Music and Forbidden Canyons, places whose very names are classics in the history of wilderness annihilation. I hike through canyonland that appears not to know cattle grazing yet. Most days I don't meet another person. The overland entrances to some Lake Powell canyons are so remote that gaining them challenges me beyond my skills or desire. Instead, I rent a boat and enter the canyons from the lake.

To escape the discomfort of visiting Lake Powell's watery graveyard, I "purify" myself of bad feelings by changing my mental image of the lake. I imagine that a volcano had erupted near its dam site, and in my visual imagery, a volcano flow, not Glen Canyon Dam, has blockaded the river. Such an occurrence is not entirely imaginary, for it evidently has transpired further downriver, at Lava Falls in Grand Canyon. Unfortunately, no volcanic flow overran the Bureau of Land Management, builders of the dam.

My mental exercises fool my Inner Nature's emotions into feeling that I am visiting a natural lake still under the self-organized care and regenerative protection of the global life system. Only then can I feel comfortable and enjoy my visit to this wilderness Auschwitz. I use the lava flow imagery almost every day. I continually need reinforcing as entering the flooded canyons brings to mind the holocaust that lies fathoms below. Glen Canyon was flooded by our public servants in Washington. It sure is hard to get good help these days.

The day that I hike Davis Gulch blasts me with an earth-shaking occurrence, making me writhe in astonishment. I return down a wild southwestern gulch. It is a Garden of Eden whose ethereal beauty fills my mind with wonders. This day I live in the splendors of an inaccessible microcosm. Its overgrown trail brings me to the deepest, most richly decorated alcoves I have ever known. Active beaver colonies, ancient Indian ruins and spectacular canyon arches bedecked with greenery touch me. Skyrocket, Scarlet Penstemon and Indian Paintbrush light the way as the brilliant sky smiles. I slosh streambeds, clamber deep arroyos and sleep in soft, warm sand. A heavenly glow fills my psyche as I return from nirvana.

But an unfamiliar Thwick—Thwuck, Thwick—Thwuck, Thwick—Thwuck, greets my ears as I near the canyon bottom. Thwick—thwuck, thwick—thwuck. I stop, listening intently. Full of curiosity and on all fours, I crawl around the next arroyo's edge. What might be calling? Louder now: THWICK—THWUCK, THWICK—ZOWIE! The shock of encountering a monster hits me. But this is for real. Before me stands Joe Golf, a middle-aged man in Bermuda shorts with knee socks, alligator polo shirt, captain's cap, golf clubs and all. Joe Golf tees off, "Thwick," his golf ball slamming against a cathedral-like canyon wall and rebounding "Thwuck," so he may "thwick" it again. HE'S PRACTICING HIS GOLF SWING IN THE HEART OF A WILDERNESS CANYON!!

I jump back, recoiling against a sand bank. What a bubble-popper! Unbelievable! As I photograph this aberration and then set up my tent, Joe Golf keeps swinging. For an hour he stands on the edge of this incredible landscape, neither entering it nor letting it enter him. Then he picks up his club, tee and red-stained ball, boards his yacht and sails off into the mainstream on the carefree highway of life which leads to that great golf course in the sky.

During my eight days at Lake Powell, I explore a dozen equally spectacular canyons and find golf balls at the foot of two of them. No question, John J. Audubon would go crazy figuring out what queer ducks laid these eggs!

Canyons: Exploring a Hole in One, *Glen Canyon, Arizona*

This whole incident hit me with much more force than did the rain of my thunderstorm encounter in Grand Canyon two decades previously. Its vigor cracked another wall in my mind, pushing through academic information and outdoor experiences that until this time had never been connected. It amalgamated five separate realizations.

1) Joe Golf is a living history museum playing out to its fullest our separated relationship with the American wilderness. Although I was once him, I'm too close to my own life to get an overview of that history. If I fully comprehend how he became the way he is, I can better understand the original source of modern people's separation from Nature, and its destructive results.

Joe Golf had thwacked me into a challenge I couldn't refuse. Uncomfortably, it interrupted many affinities. It burned within me.

2) Once on a wilderness backpack I lost my shoelaces and replaced them with nylon rope. A regular shoelace knot in the rope always untied so I devised a new knot that worked well. Another expedition participant began using nylon rope for his shoelaces. He had the same knot problem, so I taught him the knot.

During the next two years I watched others use my knot even when only their regular shoelaces came undone. They learned the tie I invented, but its source or original purpose they "new knot." They just acted without understanding why.

3) Scientists tell us that geologically speaking, humanity recently evolved in a tropical womb-like environment. That might be why we have no body hair, although most other older tropical mammals do, perhaps due to continentally drifting into the tropics. For survival in the tropics, humanity gained intense consciousness of the area's ways. That was people's major adaptive device. Early humans knew their environment's pulse.

4) Many months before a human child is born, its fetal brain develops enough to experience the human womb's chemical, nurturing, comparatively non-fluctuating environment. The fetus knows and remembers the womb as a feeling, not as an event (see **Prejudice Against Nature**, Chapter 4).

Adults may subconsciously associate protective, womb-like settings with their prenatal womb feelings. The tropics feel womb-like compared to the temperate and arctic climates.

5) Our infant survival feelings that we inherit from Earth, in prehistoric times bonded the newborn child to both its human mother

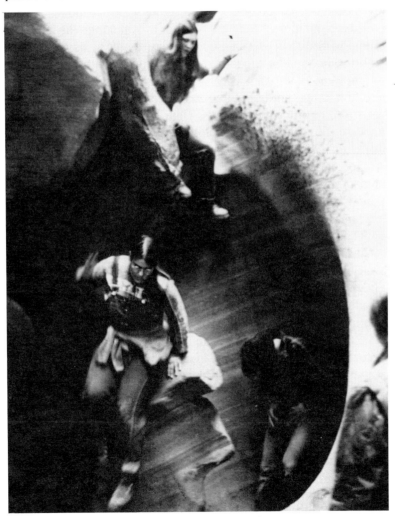

The Balance of Whole Life, *Bull Canyon Gorge, Utah*

and Earth Mother, for they were the immediate surroundings. In modern society, people-built environments and today's technological outlooks surround the newborn child. Its survival feelings bond to modern lifestyles and technologies, rather than to the natural world.

My encounter with Joe Golf bound together these five thoughts. They boiled around in my head for a week or two, and then the following purifying synthesis occurred which consistently helps me understand the reasons for modern society's conflicts with Nature.

Indigenous Societies: As early peoples migrated from the tropics into the more northerly four-seasoned environments, they experienced greater survival stress. Their hairless bodies and cultural habits had evolved to share the warm, nurturing, tropical, womb-like climate, not the food-scarce temperate winters.

Some migrating groups coped with the seasonal challenges nomadically. They moved to warmer places during the winter and to new food areas when they depleted old ones, allowing depleted food areas to renew themselves. For them, knowledge became a consciousness of the global organism's erotic motions and signals. They sensed and memorized when and where food and shelter were available. Their lives flowed with Nature's T-R fluctuations and cycles. They knew the countryside and its mannerisms like a book. They adapted themselves to Nature in the same way we adapt to the weather report. We change our behavior, not the weather.

Exemplified by the American Indian, early nomads recognized Nature's pulse as powerful, life-giving spirits: wind, rain, and landscape. They created Living Earth affinity rituals and long-lasting lifestyles spiritually connecting them to the natural world's ancient relationships. They were as one with the land. They communed with it.

Modern Societies: My modern upbringing tells me that my white race ancestors were not indigenous people but were instead of other tropical migrating groups. My ancestors pioneered modern civilization and devised a different means for survival than did the nomadic peoples. Naturally, they passed on their knowledge, and outlooks.

My ancestors seldom learned intimately from their new habitats.

They didn't flow with the seasonal temperatures and food conditions. Instead, their memories, myths and feelings kept the womb-like, relatively stable tropical environment of their origin consciously alive in their minds. And that's the outlook they've passed down to me.

Unlike their "uncivilized," "illiterate" counterparts, my ancestors technologically imposed their womb-like tropical memories on their new climatic surroundings. They changed that landscape. Instead of joining the ways of natural systems, they "civilized" them. They converted them into artificial tropics, like houses, offices, schools, cars and other technologies. And that's what I was taught to do: become literate and gain material, financial and technological security.

My ancestors' temperate-zone culture disconnected from Earth's feelingful T-R affinities and no longer respected Nature. Instead, it worshipped literate, cognitive mechanisms and attached survival feelings to cognition. For survival and security, my ancestors' womb-like tropical memories and their tropic-simulating technologies became important, not the Planet's life affinity partnerships. Artificial, people-conceived, stable conceptual schemes fortressed them against the pulse of the more fluctuating natural world. By this means they survived.

My ancestors' survival feelings made habitual their exploitation of Nature as a raw material, a resource. After that, for them to relate in any other way psychologically put them at risk, for their conceptual schemes were glued together with displaced ancient affinities and survival emotions. Today, their excessive modern descendants feel naked without the American Express Gold Card. We have more survival feelings attached to a dollar than we do to land, air and water. Did you ever try to eat, drink or breathe a dollar?

Our ancestral culture's natural affinities attached themselves to artifacts, not natural facts. Increasingly, people became cultural objects rather than children of Mother Earth. In time they learned to see the world objectively, from an object point of view. Objects meant security and survival. The Earth's pulsating life systems became grist to build artificial womb-like tropical environments anywhere—like New York's Walled-Off Astoria.

What was central for survival was these early people's tropicmaking consciousness, not Nature. It eventually affected even their religion, usually the last thing in a culture to change.

Tropicmakers' original goddess-revering Earth religions finally changed by adopting a male God, Zeus, to rule both them and Mother Nature. This sanctified their increasing technological domination of Mother Earth's "female" nature. Religiously, they could now have their will upon the virgin land.

In time tropicmakers abstracted God, made him look like them and removed him from the Planet. Their "conquer Nature" mentality fortressed Him far away in Heaven and established visiting hours on Sundays. In Heaven, He could bless their activities no matter where they lived. God was no longer attached to the land.

In this way, **male tropicmaking consciousness became God.** That's like entrusting a bear to guard the honey. If you doubt this, consider how Satan, not God, looks like Nature. With his wild animal-like ears, horns, fur, cloven hoofs, claws and a tail, he portrays Nature as evil. Tropicmakers put Hell on Earth in more ways than one. Why Satan is not female is beyond me, for Mother Nature is feminine.

Today, tropicmaking thoughts like "progress," "land improvement," "indoor aesthetics" and "conquer the wilderness," permit us to implement our tropical fantasies anywhere. We have short-circuited many of our ancient self-regulating T-R affinity contacts with Earth. As tropicmakers, we idolize our programmed minds and symbols, for we now survive by them, not by affinity relationships with the natural world. Modern messages play in our minds like movies. Our symbols are the "3 R's." We hold them as gospel.

"In the beginning there was the word and the word was God." Words are consciousness. When used to affirm comforting or discomforting feelings, they can be keys to doors that lead us to living in balance with Nature. However, they can also paint a blue wagon red, build Disneyland and explode nukes.

Fueled by our acculturated, realigned, often subconscious survival feelings, our excessive tropicmaking images and technologies destructively juggernaut. **They operate independently of the natural world's regulatory survival tensions.**

While we convert natural habitats into tropic-like areas, the natural entities and native peoples we pry or blast from their timeless survival romances often become garbage, pollutants or eradicated. Columbus came to America, planted mistletoe and kissed the Indians good-bye. Unfortunately, mainstream people who dig wilderness the most, have bulldozers.

If this sounds like outdated history, let the following news item's dateline update your awareness: "July 23, 1987. Arrows shot from the bows of Equadorian Indians today killed a Roman Catholic Priest and a Nun. The missionaries were in the jungle regions attempting to convert the aboriginal inhabitants to Christianity."

As discussed in Chapter 8, at birth and pre-birth we know only Nature's sentient T-R callings, not the ways of society; at birth, we

Feeling Society's Power, *Manahoy City, Pennsylvania*

are entirely Nature. But through our nurturing education's rewards, rejections and punishments, Western upbringing teaches the habit of demeaning our natural T-R affinity feelings and tendencies, to instead re-attach them to our culture's tropicmaking mannerisms.

All other species have cultures that enable them to survive in balance with their Mother Earth organism. We call other species' cultures "instincts" because they are not artifically conceived. They originate and connect with the Planet's survival intelligence and needs, not with the manipulation of Nature based solely on modern human consciousness.

Each of our technologies touches our natural affinities and captures them. In this process we bond with tropicmaking, becoming dependent upon tropicmaking activities and materials for the good feelings we gained directly from the Planet through Chapter 8's guided imagery. We lose the ancient, lasting Earth as a source of joy, subjecting ourselves, instead, to the runaway whims of profit, greed, technology and exploitation.

In order for each of us to help change the natural world into the tropics, modern civilization modifies our natural feelings, trains our perceptions and carves our consciousness for tropicmaking. For example, my natural hunger for food translates into dependency on agriculture, money, processed foods, packaging, stores, shopping, utensils, kitchens, stoves, fuel, dining rooms, table manners, transportation, advertising, three scheduled meals a day, and a special meaning for each food I eat. That's my food programming. These habits are as ingrained as anyone's cigarette smoking or non-metric consciousness, and just as difficult to change, too. Similarly, my excretion was toilet trained, my left-handedness restrained. But, because it destroys basic life relationships, our modern upbringing often makes about as much sense as a masochist taking aspirin.

As exemplified by our practices of circumcising an infant, traumatizing birthing or demeaning sexuality, tropicmaking often controls and changes Nature's ways, including feelings. We dismiss or give therapy to discomforting fears, anxieties and anger that we can't adapt for tropicmaking, thereby denying ourselves of their valuable life signals. We learn to disregard certain feelings to the point that, out of habit, our psyche seldom experiences them. "Are we having fun yet?" cry the anesthetized masses, 30 percent of whom obtain therapy, while an even greater percentage needs it.

In turn, our numbness to our feelings makes us excessively dependent upon tropicmaking, not Nature, for direction. That's why we smoke life-threatening cigarettes even while sensing that we shouldn't. Our survival sense is weak in comparison to the strong lifestyle irritations that urge us to smoke. Cigarettes become a fire on one end and a fool on the other.

Your state-required school attendance represents an eighteen-thousand-hour imprisonment indoors, fortressed by law from Nature's callings. Like requiring me to write right-handed, our society's artificial, left-brained, tropicmaking meanings replace our inherited whole-brained global consciousness. This makes us cause our seemingly insoluble environmental problems. From womb to tomb, **we discipline the natural world, as our Inner Nature is disciplined for tropicmaking.**

By perceiving Nature as a material resource rather than the self-organized survival relationships that we embody, **tropicmakers view Nature's far-reaching attributes prejudicially.** Our tropicmaking activities are like putting hair remover in the shampoo.

Tropicmaking causes us to exploit excessively the global organism. It makes us act as foolishly as if we considered our brains supreme and the cells of our other body organs expendable. That would lead us to mine our livers for nutrients, to raze our hair and plant our scalps with tomatoes, to use our bloodstream as a sewage system and our stomach as a toxic waste dump. Our personal survival feelings of pain prevent us from doing this to ourselves. But our tropicmaking training disengages us from having these same feelings about our Planet Mother.

We place a taboo on our natural feelings because they are wild. Yet we inherit these feelings from Earth at birth. The feeling of thirst states, "drink the water and abide by the spiritual affinity relationships and sensations that keep it available for drinking." That's the same message that every other living organism also inherits from the Planet, and why pure water is part and parcel of the natural world.

Tropicmakers need immense technological power to cultivate or replace the affinity processes of the Living Planet. At great time and expense, we teach our children that for success, most natural attributes of themselves must be extended technologically. Power creates the artificial womb of civilization, and in modern society, "good" usually means more artificial, Nature-separated or competitive. In God-like ways, we "tropically" change or subdue the natural world.

Few modern tropicmakers believe the Planet is alive, or claim the wisdom to maintain the Living Earth's formation and functions. **Yet we attempt to solve our problems with the very tropicmaking words, concepts, aesthetics and processes that caused them.**

Tropicmaking becomes runaway and excessive when our ordinarily limit-setting survival tensions and feelings from the natural world **attach themselves to our tropicmaking consciousness rather than to the living Planet's time-proven survival processes.** This culturally causes people to lose the natural restraining tensions that they have inherited from their biological Planet Mother. It cognitively disregards the natural affinity and tension controls embodied in the human brain and global life system that are essential to survival.

We must educate ourselves to once again identify, affirm and respect, rather than demean, our natural survival feelings in order to regain balance with Nature. Our environmental problems are dilemmas simply because we emphasize our artificially induced solutions rather than our limit-setting survival feelings. We must strengthen the latter and reconnect to the natural world, for it, not tropicmaking, makes survival and peace possible in the long run. Otherwise, we're driving around our delicate planet in a bulldozer, trying to resolve stress problems. I'd argue that the bulldozer we ride is the problem.

Until we discover and use holistic processes to relieve our problems of excessive artificial tropicmaking life, we continue to solve these problems artificially. It's like stupidly drilling a hole in the floor of a boat, watching water rush in and then "brilliantly" drilling another hole in the floor to drain the water out.

My encounter with Joe Golf produced the concept of tropicmaking, a unique lens that portrays modern society in a different light. Using it exposes new ways to deal with technology's impact on people and the Planet and will be discussed in the chapters that follow.

It refutes the argument that people are natural, that culture is a natural attribute of people, and therefore our tropicmaking cultural activities are natural, in turn. Experience shows that, destructively, tropicmaking is often out of contact with Nature's global community affinities and their built-in emotional T-R controls. Tropicmaking is short-circuited human consciousness alone; it excludes Gaia's collective affinity life experience, that force which is known as Nature.

Exactly fifty years after I entered elementary school, my encounter with Joe Golf helped me to understand why writing left-handed was

Engulfed By Tropicmaking, *Phosphate Mines, Florida*

taboo there. In fact, I have met people who in the 1920's had their left arm broken by school teachers because they wrote with it. You see, left-handed people are often "right-brained." They don't make good tropicmakers because they tend to learn more holistically than intellectually. They depend more upon non-verbal meanings, insights, dreams, metaphors, gestures, touching, intuition and new idea combinations than do "left-brained" people. This gives them more contact with signals from the natural world. For this reason, throughout the ages left-handed people have been considered sinister, ominous, awkward and worthless. Unlike right-handers, they at times don't respond well to the highly symbolized, verbal, abstract, logical, intellectual requirements so valued by tropicmakers.

Interestingly enough, in my particular case, I've been "tested" and come out dead center between "right- and left-brainedness." Maybe it's because I live in two worlds. It's also worth noting that "right- and left-brainedness" is not found in non-modern cultures, because, like the Planet, they're illiterate. They're more holistic, too; they believe the Earth is alive.

Excessive tropicmaking has made modern society's relationship to the Living Planet like that of a family who found themselves continually getting sick. The public health officer told them that their well might be contaminated from their septic system. So they moved the septic system to the other side of the house. But they encountered a new problem. The well dried up! I believe our present-day, chemically induced diseases and toxic waters are telling us that with respect to affinities and survival, all is not well—including the well.

DRAWING CONCLUSIONS

This chapter introduces into the global organism the cancer of modern society's exploitative, non-affinity survival. The cancer excessively converts the natural world into materials to build artificial womb-like tropical areas. The cancer's ingrained life map (consciousness) asserts that human life can best be maintained by replacing Planet Earth's global community with tropic-like indoor fortresses.

Tropicmaking accomplishes this by teaching children to abstract the Planet into objective, non-feeling symbols and to act within their framework. Society reorganizes the symbols and its members into tropicmaking schemes, then technologically injects these artificial schemes into the Living Planet's timeless affinity relationships. It has us so confused that some of us expect nuclear bomb radioactivity to cure the cancer.

The diagram below portrays our tropicmaking activities. It changes the Planet's and people's wavy blue T-R affinity relation-

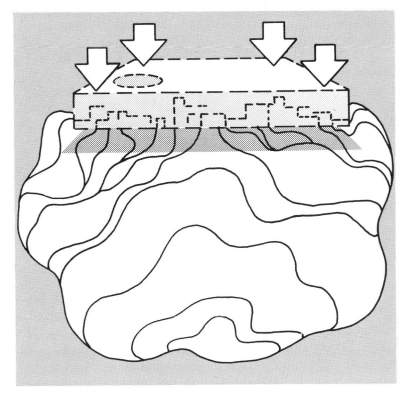

Figure 9–1

ships into abstract straight red consciousness and relationships. This separates modern people from self-organized global organism's affinity relationships with the Planet and with themselves.

In contrast to this belief system, non-tropicmaking, "uncivilized" peoples' maps are the opposite of tropicmakers' maps. Their upper level maps reflect and enjoy the Planet's timeless wisdom, not tropicmaking. Their consciousness harmonizes with blue.

STUDY GUIDE—CHAPTER 9

The author is stunned by the vast separation he views between a tourist and the wilderness.

1) Are there natural places that you wish were not destroyed by "progress?"

2) Do you find values in wilderness? What are they?

3) What experiences have you had similar to the author's meeting of "Joe Golf?" Did you learn anything from them?

4) Why do you think modern society is so disruptive, stressful and environmentally detrimental? How does your concept match with tropicmaking?

5) When have your emotions about a situation made you learn better from it something you already knew?

6) What do you think life as a hunter-gatherer was like? Compare your responses with Chapter 2 of Peter Farb's book **Man's Rise to Civilization.**

7) How do you compare the words "greed" and "tropicmaking"?

8) In what ways can you identify that your Inner Nature was trained for tropicmaking?

9) See if you can locate hidden conceptual signals and the tensions or anxieties they triggered in your Internal Planet during your acculturation. You may find this a difficult process because they are hidden. For example, how would you feel going to work naked, defecating in the street or publicly eating spaghetti with your fingers? You don't practice these taboo acts because uncomfortable tensions

would occur in you and in others. The tensions stem from the original thwarting of your Internal Planet's natural drives and feelings, as well as immediate rejection by others. Would you feel comfortable driving through a red light even if you knew that nobody was within a mile of you?

10) What, if any, taboos might exist in Nature that restrict or regulate a plant or animal?

11) In what ways do you see your daily life treating Nature as you were treated when a child?

12) Read through the newspaper. Locate ten modern symbols (thoughts, technologies, ideas) that you believe originate in people, not in Nature. Circle them in ink.

Select five natural objects or events that you appreciate. How do they meet their life needs without the use of the ten symbols or technologies? How do you feel about the positive or negative effects of the circled items upon Nature? How do the circled items affect your Internal Planet's feelings?

13) Think about five of your cultural attitudes (the 3 R's count as one). In what natural feelings and acts did this originate? How did your upbringing divert your natural feelings and/or actions to become these cultural attributes? What are your present attitudes' effects on Planet Earth?

NOTES

On this page or elsewhere, write down any thoughts or feelings that have come to you from reading this and previous chapters.

The Past Reveals the Present, *Navajo National Monument, Arizona*

CHAPTER TEN
NATURE ABANDONED

Oraibi, Arizona stands atop the miracle of a sparse yet life-giving desert. But it doesn't sparkle like a crown held high. The small pueblo village blends in with the countryside as it has for a thousand years. For Oraibi is the oldest continuously inhabited town in the United States, if not in North America.

Oraibi bears witness to an ancient way of life and the coming of the tropicmakers, five hundred years after the village's settlement. A burned church steeple stands at Oraibi's eastern edge. Its eroded form and the nearby ruts in the land gouged by huge logs, dragged sixty miles overland by enslaved Hopi Indians, remain as monuments to harsh life with the white invaders.

As we have in past years, our expedition group spends a few days with Sun Chief, a wizened Hopi elder named Don Talayesvah.[23] Although schooled in California, Sun Chief still lives out the traditional Hopi way. Because his eighty-six years have caught up with him, we help Don weed, plant and protect his farmland from the wind. It is a farmland where he grows corn, beans and squash as the Hopi have done for longer than anyone can remember. Pleased with our work, in exchange he tells us Hopi stories, sings Hopi songs and shows us the miracle of how and why his garden grows in the desert.

Our upbringing has given us a geological knowledge of the area that speaks of snow melt and underground seeps. It frowns upon his tale of this being a sacred dry oasis given by the creator to the

Hopi in order that they might always appreciate Him and his desert miracle.

From his parched garden, we hike the steep trail to Don's house. The warm interchange between us evokes a happy announcement from him: "I would like to tell you a secret of my people." He smiles, then hesitates as if awaiting some punishment for the idea. Our respectful silence urges him to continue. Don walks us out to his porch, the roof of another family living below. Engulfing us, the magnificent desert stretches out to the sky, interrupted only by buttes, mesas and long-dormant volcanic cones.

Hesitantly, with us in awe, Don looks eastward and points to a place on the horizon where a small canyon sits between two mesas. "I think I should not say this secret," he repeats. "But I want to. I am old, and they can't really do much to me," he smiles mischievously. "One of our secrets," he says, "is that each morning, here is the best place to watch the sun rise." Again he hesitates, then, still pointing to the slit canyon, he whispers, **"When the sun rises through that slit, it is time to plant beans."**

His secret seems mundane. Its meaning eludes us. Any package of seeds you buy comes with instructions on when to plant them. What's the secret? Within the week, the hidden becomes visible.

For three days, we remain in the Hopi villages. We join them in their underground religious chambers, or ceremonial kivas, and when the time is right, we watch personified, costumed kachina spirits dancing through the night to the drum's heartbeat.

Fearsome ogre kachinas drag naughty children from their homes, while their parents reaffirm the familial relationship by protecting them from the ogres and capturing them back. Guarding whipper spirits warn us away from sacred events and shrines of prayer feathers. They communicate only by screaming and chasing us with lariats.

Townspeople help us understand what we see. They relate how the secret kachina spirits affected them when they were children and how they view them now. Men who become kachinas express how it feels, what it means to them. We discover there is a kachina spirit for each aspect of the Hopi way of life: for corn, beans, bighorn sheep, baskets, et al. New kachinas appear as life proceeds. They tell us that Santa Claus is like a kachina in our lives.

People tell us that the kachinas are those who have died, returning to help those who live. The come from the high mountains that dominate the southern view from the mesa. The kachina's dance consists of stepping up toward the sun, then down toward the Earth. It vibrates between the two as does all of life on Earth.

To the Hopi people, the land is sacred. Their Gods live in it, their shrines are made of it. They come from it. They could no more destroy the land than we could destroy a cathedral by having a picnic in it: chopping up the pews for firewood, roasting hot dogs on the holy candles, placing our tablecloth upon the altar, washing the dishes in the holywater and hiring the archbishop to clean up the litter.

A Hopi priest explains that each infant has three mothers: his human mother, corn mother and Mother Earth. The child's other fathers are the sun and its uncle, who teaches the youngster the Hopi way. Very gently, without crying, the smiling newborn Hopi child, in time, is introduced to them.

Each Hopi's family includes the Earth and its spirits. It extends from the newborn to wizened grandparents. Each play an important role in parenting. The extended community, including the village, embraces the individual until death. They bury their dead in the fetal position, for they sense death to be returning to the womb of life in The Earth Mother.

As we feel what the Hopi feels, we better sense the depth of Sun Chief's secret. He and his people know Nature, Oraibi, the sun, landscape, time, community, religion, food, water, death and themselves as a single, unified act. They are mostly verbs, and they are in the present tense. Without the sun or the slit canyon marker on the landscape, they can neither grow beans nor survive. Their life secrets work nowhere else but Oraibi.

Other places have other secrets.

Hopi secrets sanctify their community life. They are their personal identity, their place on the Planet. Each secret's every word contains keys of the creator's truth for survival. That makes the Hopi cooperative organs of the global organism.

Because they know themselves as life's wholeness, their language omits the verb "to be." They are it: the peaceful ones.

The Hopi's affinity with Nature includes the Nature in people. They are a tribe of peace, unconquered, because to survive in Oraibi, their conquerors must cope with desert sparsity. To survive, their conquerors must think, feel and carry out secret Hopi affinities. Their potential conquerors neither know nor revere these secrets.

Oraibi's treasure is life, not gold.

At the end of the week, our expedition group travels eastward through the slit canyon countryside. We are dismayed to see massive drag lines leveling the hills as they are stripped for coal.

We learn that we cannot escape from sadness, for no matter in which direction we drive, the destructive impact of the mainstream's

Digging Into America, *Anasazi State Park, Utah*

tropicmaking upon Nature is inescapable. In the south, we watch the kachina's home mountains and shrines being eaten away to make roadbeds and cement. Their peaks are denuded by ever-expanding ski areas. The skiers dress much more like comic book superheroes than sacred Hopi spirits.

Westward, power plants spew fly ash, noticeably diminishing the air's clarity. Here, too, the already dammed Colorado River is threatened by new dams, even within Grand Canyon National Park. Our lifestyle appears to say "dam them all."

To the north, strip mining on Black Mesa continues, a mesa whose snowmelt naturally irrigates the Hopi desert land. Close by, radioactive uranium mine wastedumps toxify the water and air.

In desperation, traditional Hopis have appealed to modern society's government in Washington. The Hopis claim that exploitation is destroying their constitutional religious rights as American citizens. But our government has denied their appeal. It states that the Hopi's natural areas are neither cathedrals nor sacred. American laws and Bibles say these areas are resources. They are vital for maintaining modern humanity's excessive tropical urges. They indicate that downhill skiing and its gondolas uplift the American spirit.

But all is not lost, for the kachinas have found new homes in the modernized landscape; to wit: Kachina Baking Company, Kachina Motors, Kachina Heights, Kachina Boulevard, Kachina Plumbing and Heating, Kachina Insurance Agency, Kachina Shopping Plaza, Kachina Adult Books and Films, ad nauseum. Check the Flagstaff and Phoenix Yellow Pages if you don't believe this desecration! And desecration it is indeed, for how might we react to "Jesus Christ and Moses Cesspool Cleaning Corporation"? or "Pope and Rabbi Burlesque Theater"? Can you feel what the Hopi might be feeling?

Many of the workers in corporate America's exploitative sites are Native Americans. They, like ourselves, have by law attended our schools, only to have their natural affinities trained excessively into tropicmaking skills. We sit wedged in a tiny living room listening intensely as an elderly Hopi woman describes her school experience.

"They wanted us to learn white ways. They brought in the U.S. Army to make us go to school." The old Hopi woman's voice is soft as she stares at the floor and remembers her childhood. "But still my parents and the other Hopis resisted. The soldiers, every morning they came with guns, went from house to house, searching for children. So every morning, very early, one of my people would gather us children away to hide . . ."

Society in the United States has traditionally viewed the Native American as an anomaly: a race of red savages, not quite human; a feral and cunning people living in primitive conditions, close to the Earth.

One of our students of color notes that Hopi living conditions are worse than any he's seen in Harlem or down south. Like Senator Robert Kennedy, he finds the conditions deplorable.

But both he and the Senator neglect that many of the Hopi choose and cherish their pueblo community way of life because its supportive affinities feel right and have proven self-sustaining for a thousand years. That quality is lost in modern society. Other students

Hands Across Culture, *Papago Reservation, Arizona*

remind the group that we witnessed in Newfoundland the disappearance of fun-filled, old-time community dances. There, the government introduced linoleum for kitchen flooring and nobody wanted to ruin it by dancing in the kitchen as they had done for centuries.

"One by one, we were found and caught," continues the old Hopi woman. "Children and parents would be crying; the soldier would have one of your arms, your mother the other; and you were pulled away, kicking, from everything you knew, to go to the Indian school. You did not come home again for twelve years. I was one of those children."

They were Indian children, so the story went, angels of Satan, the uncivilized offspring of Nature. They had to be forcibly educated to our ways. Otherwise, "the only good Indian was a dead Indian."

"We did not have our lessons in Hopi; they were in English. Hopi was forbidden. We did not understand English, but we were punished if we were heard speaking Hopi among ourselves. The principal had a pistol full of blank cartridges. He used to fire it over the heads of the bigger boys to frighten them into obeying the rules. The food was strange to us; but if we did not eat what we were served, we went hungry. We were not allowed to go home in the summer for fear that perhaps we would not come back, so we were put to work at summer jobs. Anyway, after a few years we would not fit in at home. We were changed."

"Shades of our hostage crisis and brainwashing today," whispers one student. The living room is taut with amazed silence, broken only by the nervous laughter with which the Hopi woman punctuates every statement. She shows no resentment of the treatment she and other Hopi children had received. Her only sign of discomfort is this recurring, apologetic titter, as she looks up from the floor and reads the horror in our eyes that reflects some long-ignored part of herself and of us.

We recognize that for ourselves, although our upbringing may be different, its outcome is the same. **All modern children know more at birth about coexistence and communion with our Living Planet than they know after their high school educa-**

tion. What we call education mostly consists of placing labels on what we already know, and today's labels carry tropicmaking's bias.

The ingrained depth of the tropicmaking bias had become clear when our expedition group visited an ancient California Indian site. There we were taught by an archaeologist how to use rocks as hammers in order to chip flint into fist axes and spearpoints. For three hours, we engaged in this ancient technology. Then traveling to Death Valley, we arrived in the midst of a fierce dust storm and proceeded to set up our tents. But we had difficulty putting in the tent stakes, for the ground was extremely hard. Choking, coughing, with eyes tearing, most of the students stood in line waiting their turn to use the one hammer we had in the tool kit. Bitterly, they complained that we should have had many more hammers with us so that we would be prepared for this kind of emergency.

Yet the campground was full of rocks which would have served well for hammering tent stakes. Only four hours earlier we had even been using rocks as hammers. But our conditioned tropicmaking outlooks and upbringing had contaminated our ability to cope. We had become dependent upon a manufactured technology and called for further technological assistance to meet this situation. Out of our twenty-person group, only three thought to use rocks as hammers. A lively discussion followed in which we came to realize the pervasiveness of our programmed thinking and behavior.

Unlike ourselves, by the age of two, Hopi children's language gives them the means for knowing the Living Planet. It provides words for Nature spirits and a family including the natural world. Hopi labels and life techniques describe kinship with Nature, while modern society's labels portray techniques for conquering the land and its peoples for their resources and financial rewards.

We remembered the Hopi schoolteacher who had told us: "American education is the weapon that has finally invaded and conquered the formerly invincible Hopi, a peaceful people who once successfully repelled both Coronado and Christianity." Now we understood. And we began to question the wisdom and effects of the Peace Corps.

The old woman gestures around the room, at the three television sets, the stereo, the bookcases full of books, the couch, the electric dryer, the coffee table, the wall-to-wall carpet. "Now I am an American," she says proudly. But there is a crease between her brows and she laughs away her stress.

The expedition comes away from the Hopiland visit having sensed the joys and sorrows of the southwest. The joys emanate from people and the natural environment mutually growing and supporting each other. The sorrows stem from contact with tropicmaking's harmful impact on the land, on the Hopis, on ourselves.

Although we feel bad about the situation, tropicmaking teaches us that our feelings have no intrinsic value. We feel helpless and at times apathetic because our traditional education has given us advanced courses in shrugging.

Our Hopiland friends show us their sadness. As they gain the skills of tropicmaking, they reap the hurt from breaking the timeless affinity bonds between the Planet and themselves. In the villages, drugs, alcoholism, stress, poverty, divorce, apathy, suicide, violence, crime and mental illness flourish where before they were unknown. Some say the Bureau of Indian Affairs is part of the problem. BIA is said to mean Bust Indian Asses.

We come to recognize that the Hopi's demise replicates our own plight. At a living history museum in Massachusetts, we role play being Americans living in the 1820s. Then, most American families lived in farm communities as do the Amish today. But the mills lured farm families into factory work, and the shrill scheduled sound of the mill whistle replaced the gentler life-pulse callings of the farmland. People became extensions of the machine world, and education became a tool to make it happen. Divorce increased as survival no longer held people together; survival could be bought with dollars.

We begin to see that the Living Earth's physiology includes all people's inner biology and emotions. **As we excessively attack natural systems to create the likes of the tropics, we also excessively attack the Nature within us.** The Bible recognizes it in Ecclesiastes 3:19: "For that which befalleth the sons of men befalleth beasts; as the one dieth so dieth the other."

Our attacked Inner Nature responds to its abandonment with aggression, anxiety, depression and the destructive dependencies associated with them. In this way, modern life often becomes like a cesspool; what you get out of it depends upon what you put in it.

As we habitually think in our exploitative modern ways, our Nature-abandoning ideas continually attack our Inner Nature, as they do the Hopi. No matter where we are, that is the hidden source of modern psychological stress. It's not only outside; we carry it with us. We live in the anxiety of wanting to extend our hands to give and receive love, yet not knowing whether the environment will greet or slap our desires. Too often a modern person extending a hand toward another is viewed as a purse snatcher, or worse.

Excessively, therefore, modern people must put themselves at risk to express their true nature while low self-esteem consists of the words of inadequacy we learn to apply to our Inner Nature's difficulties in handling tropicmaking skills. The discrepency between our artificially induced "mind movies of consciousness" and the way Nature works causes most of our mental ills and anxieties. We note

1985 Amish Mustang Convertible, *New Holland, Pennsylvania*

that experimentally, the neurosis seen in laboratory and zoo animals is seldom found when they are in their natural habitats.

Tropicmaking castrates us of our ancient life relationships with the wilderness. We lose contact with powerful, worldly forces of the bear, forest and mountain. For them we substitute people-contrived technologies whose operation is limited by the warp of our tropicmaking opinions of Nature's purpose as a raw material.

As long as objective science, or anything else we use, negates our Inner Nature, it becomes part of the stress problem. Even outdoor education's artificial stress settings, which give people social confidence by conquering Nature, are misguided. They don't fully recognize that our disorders stem from our Inner and Outer Nature already being hurtfully conquered. **That's the stress we must label and address.** Without identifying this truth, we overlook it to our cost.

Our Hopi visit showed that the Earth is home for the Hopi and for us. It allowed us to see that other species and other cultures know the secret for living in the biosphere, the secret of Nature's affinity relationships and T-R survival sensations. Without using reading, writing or arithmetic, the Hopi survive better than we do. They have learned to think with their hearts. It is these relationships which contain the wise, powerful, homeostatic forces that maintain Earth's T-R harmony. Our artificial upbringing amputates them from us and the global organism.

As emotional amputees, when we look in a mirror, we see a zombie as our reflection. That's different than seeing our reflection in the water. There it ripples alive with the pondlife and wind spirit's pulsations. Without our wilderness spirit, we are mere ghosts of our former selves. Today we spend much of our daily lives seeking sensations of some kind to replace those lost by our emotional separation from Earth. In modern society, sensationalism and winning become pervasive simply because they help us know that we're alive.

We seldom recognize that horror movies only work because they bring to consciousness the long-suppressed, frightened feelings from our upbringing, including trauma from unnatural birthing practices.

Seeking Their Own Truths, *Martha's Vineyard, Massachusetts*

Our Inner Nature recognizes these films' terrors because **we have lived them on some level during our tropicmaking acculturation.** The films make our subconscious backgrounds explode. Yet, they do have their sentimental moments, too—like when lightning strikes across the clouds and Frankenstein cries out, "Mother!"

While living in the natural world, I've designed the following true-false test. It confronts, and then helps explain how some of our common knowledge makes us zombies by disowning us from Nature. It's a secret well worth knowing. Take the test and ask yourself why the following cultural truths that you carry in your consciousness are false in Nature.

1. ( ) True (x) False	$1 + 1 = 2$
2. ( ) True (x) False	The shortest distance between two points is a straight line.
3. ( ) True (x) False	An apple always falls down because of the law of gravity.
4. ( ) True (x) False	Gases, liquids and solids act predictably under standard conditions of temperature, pressure, etc.
5. ( ) True (x) False	$I = E/R$, $S = \frac{1}{2}gt^2$, $Ca + 2HOH = Ca(OH)_2 + H_2$, $V^2 = 2as$.
6. ( ) True (x) False	Columbus discovered America.
7. ( ) True (x) False	The Earth's diameter is approximately 5,730 nautical miles.
8. ( ) True (x) False	Geographically, people live on the face of the Planet, not inside.
9. ( ) True (x) False	The bulldozing done to build a large shopping mall is similar to the "bulldozing" done by a glacier's advance and retreat.
10. ( ) True (x) False	Nature is governed and can be known by stable facts, figures, laws and cycles.

Life consists of Nature's healthy, fluctuating T-R pulse. Rather than to flow with the rhythms of Nature, our acculturation teaches us written, static laws that ignore Nature's ever-changing moods. Often when we apply these laws, we stop Nature's pulse. We divide the whole of life relationships. The entities that permanently lose their natural life partners are called pollution, toxic wastes, chemical contamination, acid rain, radioactivity, nuclear holocaust, social tensions, war, emotional stress, ad nauseum.

From the viewpoint of natural processes, then, consider these fallacies in our Western consciousness:

1. $1 + 1 = 2$? No. **For years, I've searched for "one" in Nature.**

I can't find it. In Nature, over time, "one" doesn't exist; "one" always changes. For example, **one tree is never exactly like another tree,** and one tree **always differs from moment to moment**. Like us, it is always in a state of growth and adaptation to its ever-changing environment of wind, water, animals, plants and minerals. "One" is a cultural constant, but **in Nature the only long-term constant is change.** Mathematics, statistics, economics and sciences solely based on the truth of "one" are false. The closest Nature gets to "one" is wholeness—one ever-changing Planet Earth or one universal expression of affinity. And, since Nature exists everywhere, "Zero"—signifying nothingness—can't accurately represent the natural world either.

2. Few, if any, straight lines exist in Nature, and those that do, change in time. When applied, straight lines and straight-line thinking distorts, stresses or destroys natural systems. Most geometric axioms are "given"; but they're seldom, if ever, true in Nature. They just feel right to us because of our upbringing.

3. Many laws of physics are unholistic because they subdivide Nature. If the law of gravity always causes an apple to fall, how does this law explain an apple rising to get up in the tree, which it does year after year? The fluctuating rising and falling of apples, and everything else, is the rule in Nature, not gravity alone.

4. Matter does act predictably under standard conditions of temperature, pressure, wind velocity, etc. But only in laboratories, not in Nature, do standard conditions exist. To make laboratories in which matter acts predictably, people must manipulate and control the pulse of Nature. That causes the negative environmental impact of so many technologies.

5. As long as people who apply formulas are not held responsible for their accuracy, wholeness or adverse effects upon the Planet, the formulas will remain as apparent truths. Because most formulas or equations negate Nature's ways, their positive potential is often negated by their environmental impact. For example, calcium doesn't exist, as such, in Nature. It's always combined with something else. And gravity and time are known to fluctuate.

In the long run, most value judgments and trade-offs with natural systems compromise the Planet's life. They subdivide the whole of the Planet's life, which results in its deterioration.

6. North America existed and was inhabited by Native Americans for many thousands of years before Columbus was even a twinkle in Queen Isabella's eye. Western culture tends to demean Native Americans, women, people of color and others because they appear to live closer to Nature's ways and callings.

7. Planet Earth does not only consist of what we see of it. It's diameter invisibly goes thousands of miles into space. The Planet includes its atmosphere, electromagnetic wave bands, gravity force and relationship to the sun.

8. People live inside, not on, Planet Earth. We and our homes lie imbedded in the Planet's flesh and blood of air, water, soil, wind, rocks, atmosphere, gravity and force field.

9. The forces guiding and empowering a glacier are knowledgeable, ancient, self-organized, regenerative, life experiences of Gaia in response to its T-R relationship with the sun. In contrast, the bulldozer's design and operation is guided by the excessive artificial-environment desires of tropicmakers who act from survival feelings that, unguided by Gaia, are attached to economic gain.

10. In order to survive as a living organism, Planet Earth often reacts as unpredictably as a wild bear to the environmental factors that challenge its survival.

I include here the following story because it humorously catches the conflicts that stem from our abstract communication, misunderstood technologies and disrespect for natural functions.

Mrs. McCarthy visited her doctor who requested she bring him a urine specimen first thing in the morning. The next day she appeared at his office scratched and bruised with one eye quite swollen.

"Goodness, what happened to you?!" asked the physician.

"Well, Doctor, to tell you the truth, I didn't know what a urine specimen was, so I went to my friend Mrs. Flaherty and asked her."

"Go piss in a bottle," says she.

"GO SHIT IN YOUR HAT," says I, "AND THAT'S HOW IT ALL STARTED."

As with the Hopi, so also with us, it is essential that our symbolizations reunite our Inner Nature with our Earth home and with each other. Only when the symbols of feelings, Nature and research synchronize does our mentality allow us to touch the Earth's wisdom, enjoining us with life's regenerative powers.

Many people resent this notion. They believe that being environmentally responsible is a form of fascism. Preserving the Earth's life would limit people's freedom to over-extensively develop or exploit their environment. This may be true, but isn't our health and survival, and that of our children, more important than selfishly, greedily having more than we need all the time? A recent study shows that, **be they extremely rich or very poor, most people feel they need approximately 20 percent more money than they presently receive.**

Democracy, especially by consensus, is the best form of government because it mimics the Planet's governance. It lets each person's T-R affinity comforts and discomforts be known and accommodated by all. But **even a democratic state is self-destructive if it operates outside the Earth's life system.** Then it can't "hear" the Planet's life and react to it. Then society becomes like a street gang, a smoothly operating but estranged and uncontrollable cancer of its environment and finally of itself. A democracy by consensus reduces this problem because it gives a platform for natural feelings, for their source and their rationale.

I believe Sun Chief had the right idea. He concluded, "As soon as everybody loves the land, we'll all be very Hopi." [Happy]

To Love the Land: Sun Chief, *1976, Oraibi, Arizona*

DRAWING CONCLUSIONS
By using colored arrows, the map portrays the conversion of natural, blue life-affinity global and personal relationships into red tropicmaking artifacts and desires.

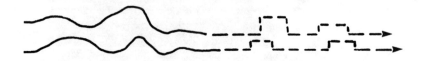

Figure 10-1

STUDY GUIDE—CHAPTER 10

1) What aspects of Native American culture have you learned to mistrust? What values might these aspects contain?

2) What "life secrets" do you believe you have been taught?

3) What guiding Nature spirits, if any, do you recognize in your life?

4) Would you consider St. Francis of Assisi to be a natural spirit? How about the Lorax?[7]

5) How do Hopi secrets protect the land from excessive exploitation?

6) Do you believe that our modern truancy laws are the same as mandatory education for the Hopi?

7) Has your education removed you from Nature? Has it changed you so that you would no longer fit into a natural community?

8) What evidence do you see for or against the author's notion that "the Hopi's demise replicates our own plight?"

9) Where have you experienced stress from the abandonment of your Inner Nature?

10) What experiences have you had which show that chemicals harmful to the natural environment can't be tolerated by people's biology?

11) Select five important machines in your life upon which you depend. What services do they provide you?

Select five natural objects. In turn permit yourself to imagine yourself being them. As each of these natural objects, how do you satisfy the survival needs that the machines satisfied for you?

12) Permit yourself to become at one with a cultural object and then a natural object. What differences do you feel between the two? What similarities? Say them aloud. Can you find an internal Planet in a cultural object? In a desk, car or gun? Is that its spirit?

13) Think about four examples in your own life where your assignment of meaning, value and importance modifies natural life systems. Has your internal Planet been stressed or impacted while achieving Western culture's meanings, values and rewards? Are your personal meanings and values biased against Nature and your internal Planet?

14) Name six technologies you use that separate you from Nature. What steps might you take to bridge this gap? What prevents you from doing so?

NOTES

On this page or elsewhere, write down any thoughts or feelings that have come to you from reading this and previous chapters.

Mapping Our Common History, *Navajoland, Utah*

CHAPTER ELEVEN
RECHARTING THE COURSE

Modern society's inaccurate life map is not just a matter of embarrassment. It's a matter of survival and a matter of time.

We know our life map is wrong because the more we use it, the more lost we often get with respect to living peacefully with the environment and each other. Our life map short-circuits our rationale. We make war to have peace; we need more nuclear weapons to have less weapons. We must feel stressed to feel good, knock others down to build ourselves up, conquer Nature to gain self-confidence and feel pain to gain. The last decade's technological miracles cause today's problems. Although bathed in time-saving devices, our modern lives are stressfully hurried, and the electric company tells us not to use the appliances they tell us to buy.

The most serious map problem is that **our map doesn't have marked on it the more harmonious place where we want to go.** Using our present map, therefore, try as we may, life can become more stressful. For example, writing left-handed was not on the map that was used during my childhood. That's why the map gave problems to my teachers, my parents and myself, and vice versa.

My students and I use the revised whole life map this chapter presents. We know it works for it recognizes examples of how to attain the harmony we seek and puts them on the map. To revise our old map, we must give it the depth necessary to produce the harmony we desire. It must have the survival value of whole life, for survival is the bottom line with respect to life. Survival must appear on the map. It obviously isn't there now. Far too many people and natural systems are unnecessarily stressed, starving, diseased and dying, both at home and on the battlefield. Putting our whole life support system on a map prevents us from overlooking it.

An Overview: The Ecology of our Inner Nature

We survive because we share with the Earth the basic life necessities of food, water, air, shelter, climate, temperature and affinities for them. At the time humanity made its appearance, the Earth of itself in the preceding six billion years had already organized its entities through affinity relationships. They maintain and regenerate life as we know it in its natural state. They have a strong union and a full-time subcontract with God. He rests on the seventh day while they keep things going.

People did not design the map that maintains life harmoniously, nor is it a written map. The Earth's and people's biological map is the global life system's ongoing, mutually beneficial community experience. It is affinity relationships, not a chart, and affinity relationships are carried in the heart, not the glove compartment.

Each species has its own special gifts for surviving. Humanity's special gift is intensified consciousness and toolmaking. That means intelligently using symbols and images to engage us in the Earth's relationships. To be harmonious, our map of consciousness must accurately convey ourselves as an embodiment of the Earth's relationships and make us relate to them. If it doesn't induce mutually beneficial relationships, the map misguides us as it reinforces our cognitive separation from Nature.

Like any other society or culture, modern society's survival consciousness maps out how it thinks the world works and how it thinks we ought to relate to it. Modern society habitually enacts the symbols and images of a tropicmaking map.

The Person-Planet Map: A Unified Whole Life Picture

Harmonious survival means becoming not a separate but an integrated part of the whole of life. For this reason, the map presented

Bring the Word Outside, *Superstition Wilderness, Arizona*

in this chapter depicts two things at once. It is a map both of a person and of the Planet. Biologically, it unites them. Any reading of this map portrays what is happening both on the Planet and in a person. **The map prevents separation by not being separated.**

However, even with this map we must be aware that, by itself, any map or book is misleading, in that it is not experiential. It is merely an abstract cognitive tropicmaking device limited to one dimension: symbols and images. Our actual world consists of many dimensions: up, down, horizontal, time, symbolism, sensations, feeling, thoughts, acts and relationships. Although this map contains neither feelings, actions nor relationships, it can help you discover the survival value of all of them.

For the map to accomplish its goal, you must give yourself permission to involve yourself in a map whose every figure represents both people and the Planet. The map's imagery reflects the material presented in the previous chapters, but it also represents an inner desire to beneficially unify yourself with people and the Planet. **Without your desire and consent to do this, the map will be useless to you.** The choice is yours. It should be made before reading further. One way of consummating this choice is to formally contract with yourself to understand a unifying person-Planet map. Although it may feel foolish, you may do this by signing the following statement.

I, the undersigned person, recognize and desire the benefits of a less separated relationship with Nature.
I consent to be involved with constructing a map whose every symbol represents both a person and Planet Earth.
Date _____ Signature _____
Witness (optional, but reinforcing) _____

To proceed, you will need blue, red and yellow pencils, crayons or pens.

Figure 11-1

We start with the blank space of figure 11-1.

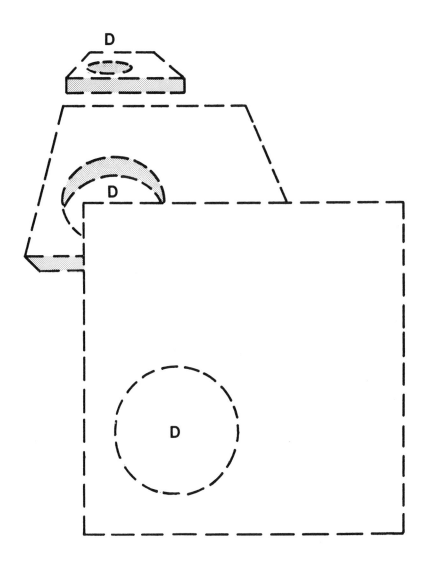

Figure 11-2

I draw the map's space in Figure 11-2.

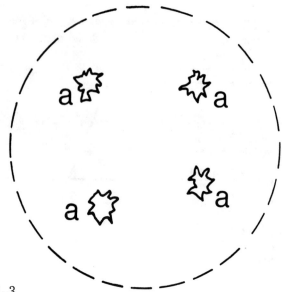

Figure 11-3

In Figure 11-3, I add some markings on the map. Color them blue. What do you see in Figure 11-3? At my workshops when studying Figure 11-3, people say they see stars, jiggles, marks, blue particles, butterflies or entities. Rarely, if ever, do they see the air between themselves and the page, the space between the blue entities, or the design and sensations of the total environment around them.

As tropicmakers, our upbringing trains us to see material objects to which our attention is directed, but not to see the space or relationships between them. To be aware of this deficiency, the map brings it to consciousness by using wavy connectors in Figure 11-4. Color the entities and connectors blue. These connectors represent six aspects of any entity which our upbringing usually hides from our awareness. Again, these entities and relationships exist in people and the Planet.

   1) Each entity is in motion.
   2) Each entity has, or perhaps is, an affinity to relate to other

Figure 11-4

entities. Motions and emotions may be identical.

3) Each relationship vibrates and fluctuates from tension production and tension release (T-R).

4) Each established relationship can be identified as a new, more stable entity, which, in turn, has an affinity to relate to other entities.

5) Entities in actuality may not exist. They may only be the energies or affinity feelings that we can identify, or to which we can relate and apply labels. By labeling them, we bring them onto our symbol map of consciousness.

6) Since we are typical entities and contain a consciousness map, on some level every other entity does, too. Each entity can somewhat determine how it will respond to signals.

Items 1-6 are universal facts of life. With respect to our solar system, Planet Earth and ourselves, geological evidence indicates that planetary relationship building between entities in this part of the universe has been taking place for the past five to seven billion years.

Most people believe that a circle is the best way to diagram Planet Earth today. It is not; it is merely the easiest to draw. A circle only accurately images the mineral, "inert" portions of the Planet. This is the imperfectly circular portion that our diameter and circumference measurements allude to, and measurements seldom include the atmosphere, troposphere, ionosphere and biosphere. Actually, the mineral portion of the Planet is slightly elliptical due to its relationship to the sun. A circle doesn't portray that, nor does it portray the Earth's and/or a person's fluctuating T-R life relationships internally with the sun, solar wind and universe.

Living in Nature leads me to diagram Planet Earth and people as Figure 11-5, an amoeba-like form. Color it blue. Its wavy circumference represents its many T-R affinity pulsations and the atmosphere's cell membrane quality. Its circle containing stars represents the blue entities and recycling connector relationships discussed in Figure 11-4. Color them blue. They produce the wavy T-R lines.

If we actually photographed the Earth from space with respect

Figure 11-5

to the solar wind, the Planet would appear as a wriggling, comet-like tadpole, or sperm, swimming toward the sun.[18]

Figure 11-5 depicts the Planet and people before the advent of modern society. Its blue color and T-R fluctuations state that biologically and emotionally, people have been part of the Planet's growth since its inception. The blue arrow portrays that people and the Planet expressed their T-R needs and drives naturally, in cultural and biological harmony. In congress, over billions of years, their sensation-networked nature organized, perpetuated, regenerated and evolved into the living organism, Earth.

About Spirituality

As previously discussed, Planet Earth, like a giant, warm-blooded plant cell, has an Earth-edited respiratory or spiritual relationship with the sun. Figure 11-6 portrays this scientific, spiritual relationship.

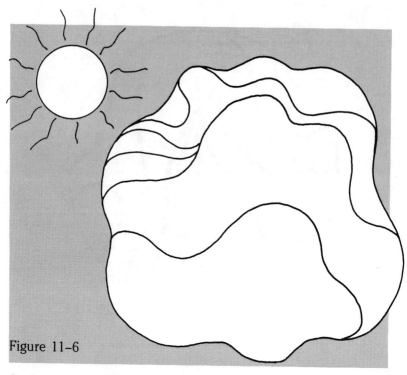

Figure 11-6

Background for Figure 11-7. As described in Chapter 9, some early peoples found that they could survive by changing the people-inhospitable aspects of the natural world into artificial tropic-like living areas. They didn't survive by honing their consciousness of T-R sensations of Nature. They depended instead upon intellectual symbolization techniques that used the natural world as a resource for building tropic-like survival pockets in temperate areas.

Uninterrupted by Nature's sentient callings, the tropicmakers' abstract survival map consists of symbols and images. Indeed, tropicmakers usually value intellect more than feelings. In our tropicmaking society, feelings have no survival value until they are converted into tropicmaking techniques. Tropicmakers become overly conscious of artificial feelings—fashion, power, education, status and wealth—but not of survival with Nature. Subconsciously they attach their survival feelings to tropicmaking manifestations.

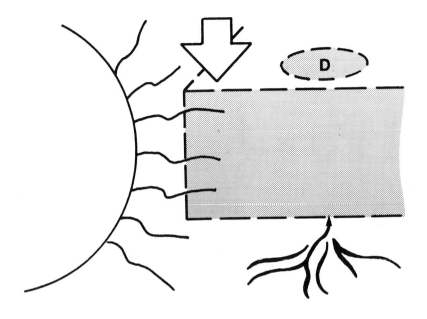

Figure 11-7

Part of the tropicmaker's consciousness map conveys that other cultures survived by using maps honoring the sentient callings of the global life community. To tropicmakers, these cultures appear uncivilized, close to Nature. As tropicmakers exploit Nature, they exploit these cultures. That is the nature of tropicmaking.

Every aspect of the previous Figure 11-6 is grist for tropicmakers' mills, including the natural physiology and emotions within themselves and others. Historically, "I feel affinity, therefore I survive," in time becomes "I think, therefore I am."

Figure 11-7 places tropicmaking's artificial means of survival prominently on the Planet-people map and represents the processes and impact of modern life. Color all areas red within the dotted lines.

1) Tropicmaking is colored eye-catching red to distinguish its artificial people-manufactured products and ways from the blue-colored natural community's ways.

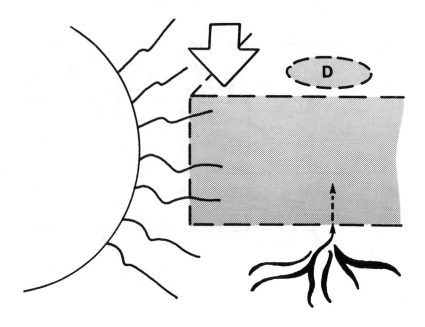

Figure 11-8

2) It is constructed of straight lines to depict modern, single cause and effect thinking, and the halting of the natural wavy T-R life pulse.

3) Its square-cornered, rectangular shape mimics the shape of modern people-built construction, technology's black box, a jail with limited routines, and the dollar, a metaphysical symbol empowering its owner with control over Nature and people.

4) Its thick walls represent modern society's desire to separate itself from Nature by using natural materials and power to construct tropic-like fortresses for people.

5) It has growth arrows representing its cancerlike survival desire for progress and economic growth that further exploit natural systems and pry or blast apart timeless natural affinity relationships.

In Figure 11-8, the arrow changing from wavy blue to straight, dotted red shows how natural systems and feelings in people and the Planet are captured and converted by modern society's ingrained

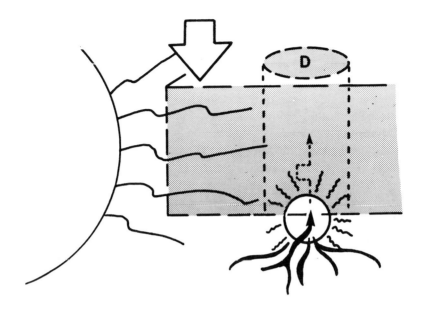

Figure 11-9

tropicmaking consciousness. In modern society, blue needs, relationships and survival feelings are attached to "artificial safety fortresses," like power, houses and schools, that separate people from Nature's fluctuations. We are "toilet trained" into tropicmaking.

1) The arrow shows that modern people habitually remove from Nature the blue affinity relationships that produce natural materials. These relationships are shaped by people into red artificial objects and tropic-like environments.

2) The blue natural feelings, thoughts and actions within people are lulled, goaded and trained to be experienced as red tropicmaking desires, skills and gratifications.

3) The red arrow is fueled and moved by powerful, often subconscious, natural survival feelings that in Nature are consummated through the establishment of mutually supportive affinity relationships.

Figure 11-9 goes back to basics. It introduces to people and to the Planet the life-giving respiratory energies and light of the sun at the point where the blue arrow begins to change into red. Color it yellow. This "spirit" light area colored sun-gold yellow indicates that it:

1) originates in the universe and sun which produced the solar system and the Planet's affinity for mutually life-supportive relationships.

2) needs space in order to function.

3) needs time in order to function.

4) sheds light and energy upon whatever it touches.

5) is as foreign to our everyday consciousness as would be a foreign language.

6) is an essential confronting area of the map if natural systems and sensations are to cease being habitually converted into red artificial tropical environments, feelings or relationships.

7) is the time and place to introduce the whole revised map onto a modern person's habitual tropicmaking map. It is the point D discussed in Chapter 2.

8) makes the map holistic by introducing a time and place to become conscious and consider thoughts, feelings and actions.

Figure 11-10 introduces a new dotted arrow that returns us to our origins in order to reconsider our relationships and direction. Symbolically, it uses the natural light of the sun to illuminate new ways of relating. It allows us to attach our survival feelings to the Living Planet's needs and Nature's ways instead of tropicmaking's destructiveness.

New relationships are formed by choosing to blend the golden sunlight of consciousness with the blueness of Nature's affinity ways. This blending of blue and gold produces a new, green direction that becomes an alternative choice for modern people. Color the dotted arrow green by mixing blue and yellow. Sometimes you must rub them with your finger, hands-on, smudging them together to get green. Do it. Don't be afraid. The new arrow, dotted, denotes a cultural hands-on consciousness that harmonizes with the Nature and affinities found in people and the Planet. It makes survival a function

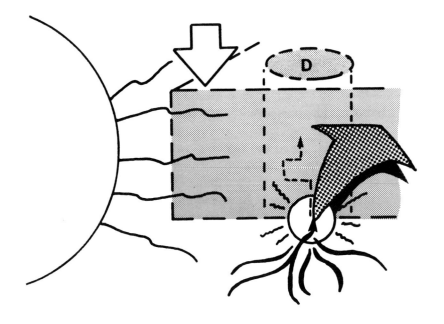

Figure 11-10

of healthy, natural systems. You develop life ways that encourage natural feelings and allow systems to grow.

When the yellow sunlight touches tropicmaking red areas, it produces an orange color, denoting danger, like a caution light. For long-term health and survival, we must regulate these relationships. They usually inject stress and toxins into life systems and excessive carbon dioxide into the atmosphere.

In modern society, we best attain the green direction by creating sunlit hands-on relationships. Their basis is awareness of how the Planet functions as a living organism. Green relationships make their participants become organs of the Living Planet. They are worthwhile because they feel right for all parties. They join people's Inner Nature with its biological mother, the Living Planet.

Green relationships fuel themselves on the joys that one finds in the nature of people and the Planet. They're more than just hitting yourself on the head with a hammer because it feels so good when you stop.

* * * * *

To use the whole life map, you must:

1) provide for yourself a time and place that allows you to study the map and become involved with your Inner Nature, comfort and discomfort feelings and the golden glow area.

2) allow yourself to be confronted or in other ways discover and validate how you feel about anything.

3) one at a time, like peeling apart an onion held together with emotions, find the blue natural origins of your red tropicmaking feelings.

4) choose a greener, hands-on, more holistically satisfying direction for your blue feelings' expression. If you need some green ideas for yourself, read John Lobell's **Little Green Book.**[13]

5) build habit-changing relationships allowing 1-4 to happen.

6) repeat the whole 1-5 process, as time continues, until it becomes habit.

Remember, learning to use the map often takes longer than learning a new language. It involves learning to find and trust your feelings and their messages with respect to short- and long-term survival. Besides making life more healthy and harmonious, the map is also practical. In time it makes you green enough to get a full-time job as a forest, or as a jolly giant who grows vegetables (just kidding!). No matter what you do, from your efforts the environment and you will feel better and survive more cooperatively.

At some conference workshops we leave our unusual map sitting on the blackboard. Workshops on other subjects later use the room and report that they don't understand the diagram we left. Lately, as an experiment, I've been labeling it: Miami Dolphins (blue section), Washington Redskins (red section). Diagram features receive labels like nose guard, fullback or tight end. The green arrow goes off tackle. We leave the room wondering what the next group will think we discussed. What's amazing is that nobody has questioned the diagram since I've labeled it as a football game.

Blending Blue and Gold, *Grand Canyon, Arizona*

STUDY GUIDE—CHAPTER 11
A map is created that joins people and the Planet as a single organism.

1) From your own life, find examples of misusing a physical map and having your internal map mislead you.

2) What differences and similarities do you see between your society's life map and the life map used by the Living Planet?

3) What, if any, changes would you make in society's map? For what reasons?

4) Design a map that you believe is more accurate with respect to Nature and people.

5) Do you think you have ever been involved in the golden glow area?

6) Where do you see places in your personal life that the golden glow area could be established?

7) Try to identify the different signals to which your consciousness responds during the day. See if you can identify survival sensations and feelings emanating from within you, as well as signals from your surroundings.

8) Observe or imagine a living plant or animal. Identify or imagine the different survival sensations or feelings it may experience on some level.

9) People gain security by trusting their ability to act for Internal Planet gratification and against discomfort. Try to find examples of this in your life and in the lives of others. Say aloud or write where you have gained or could gain security from your Inner Nature by choosing gratifying symbols or situations. Watch a wild animal or plant for a half-hour. Imitate the actions it repeats. Explain how it gains security (survival) from these acts. Apply your explanation to aspects of yourself that give you security.

10) Whenever you feel personal stress, hurt or depression, or see it in others, try to discover how and why the individual's Nature

is experiencing abandonment. Do this with the feelings in dreams and fantasies as well as with conscious life. Then design words and deeds to protect Nature from any undue abandonment.

NOTES
On this page or elsewhere, write down any thoughts or feelings that have come to you from reading this and previous chapters.

The Whole Earth was of One Speech, *Everglades National Park, Florida*

CHAPTER TWELVE
THE GOLDEN GLOW

Tropicmaking is a form of bigotry. It is based on the idiotic notion that the natural world must die for people to live.

Fully-trained tropicmakers are repulsed by the idea that Nature should be respected, not conquered. I understand their shock. Before I put on my hiking boots, I stood in their shoes.

Confirmed tropicmakers usually tell me that you have to prove the person-Planet map is a fact. If you can't scientifically quantify or qualify your notions, they're useless. That reaction reminds me of the confused motorist who found himself on an unmapped dirt road in the mountains. "What's this road's name? Where does it go? How do you get to the city? Which way is north?" he asked a smiling farm boy who couldn't answer the questions. Aggravated, the motorist commented, "You sure don't know much."

The boy mused, "Maybe so, but I ain't lost."

The global life system map may not answer tropicmakers' questions scientifically, but with respect to life, it's not lost.

Planet Earth communicates through touch-feel signals and relationships that are taboo to many tropicmakers. The map shares the Planet's design. Given a chance, it puts anybody more in touch with it, through their feelings. Its bottom line is simply whether or not a person trusts his/her comforting or discomforting (T-R) natural feelings.

Any map of consciousness is part of a person's culture. It is a living portion of the person who uses it. That's why each of us is prone to becoming aggravated or defensive by increased exposure to new or different ideas. They confront our acculturated personality traits and trigger hidden "you're wrong, bad or stupid" feelings which have been attached to our Nature by our upbringing.

Although I've used the Chapter 11 whole life map nationwide for years, recently the map most validated itself for me in, of all places, a dream. I trust the feeling messages dreams can give because I don't control them. Some of us also trust in God for the same reason.

As usual, I was sleeping on the ground, this time on the volcanic cinders of Arizona's San Francisco Peaks. In my dream, I was allowed to experience the blue part of the map. The red portion was there, but the golden glow of consciousness allowed me, for a very short time, to not let the red part touch me. I was all blue.

The powerful, warm, secure, joyful, enlightened feeling of those moments of blueness conveyed how unadulterated life felt. That incredibly elated dream and powerful feeling were either like being an embryo in a very supportive womb, or like being dead. Either way, that feeling affirmed for me what this book's mapping of experiences attempts to communicate with symbols.

In their own way, the dream feelings stated, "Stick with the map. The worst that will happen is that it will kill you and return you to Nature, the womb of life." **Well, if death feels like that dream, then it's something else to look forward to.**

I've often wanted to give the map and the dream's powerful feelings to friends that they might enjoy its truth and security. I wonder if hypnotized people, familiar with the map, could be allowed to remove their redness and reassociate their good feelings with their blue Inner Nature and its mother Planet. Once might be enough to induce welcome cultural changes in our society's self-destructive areas.

The feeling and the map give me confidence. I can explore anything I choose to, so long as I enjoy the feeling of the Planet's wilderness within me . . .

* * * * *

The whole life map's bright space of golden sunlight only exists in the present moment. It makes the map useful because in the NOW of life, you can put things together and do something about situations. The NOW is Nature. It is where you can make cultural changes leading to more balanced relationships with Nature. The map allows you more harmony based on your personal outlooks, history, needs, feelings and abilities. You become an important vehicle for cultural change. Don't wait for the government to make changes. It won't until you do, for the government represents you. Don't wait for God either; God is the last thing to change, for It is the sum total of everything, including the government.

How different that type of thinking is from the tropicmaking mind games about past and future. They divert our energies away from immediate, feelingful relationships.

Apathy reigns when we've lost feelings of motivation. Then the only thing we can do fast is get tired. If we desire the golden area,

Immersion in Feelings' Origins, *Great Sand Dunes, Colorado*

we build it because we value it. WE produce it by sustaining relationships that allow us the golden area's time, space, map and meaning. Often it is a meditative period, one where we heed the map and seek signals from our blue Nature that is buried under the artificial red symbols and feelings of our tropicmaking upbringing. The golden area demonstrates deep ecology. It is modern society remaining home with the natural world.

Most modern problems exist because we get enjoyment and encouragement for relating in ways that cause them. They are habit. Information alone doesn't change habits. The golden glow provides time and space to bring this to our programmed consciousness. It lets us regenerate Earth affinity relationships. It provides the opportunity to create new habits, to enjoy life in helpful, non-destructive ways. Even Santa Claus isn't doing that right now. Too often he builds warlike toys whose materials pollute natural systems and children's minds.

As tropicmakers, we gain support by becoming artificially inclined, by abandoning or attacking Nature. Modern society doesn't accept us without our profitable tropicmaking habits, skills, aesthetics and competencies, such as engineering, construction, forestry, mining, language, teaching, art, etc. A deep understanding of our feelings, values and sentient relationship to wilderness is neither required nor taught. The apex of modern life is to be so aesthetically or technologically encumbered that it is difficult to differentiate yourself from cultural objects and exploitation.

Most personal problems exist because people's Inner Nature is thwarted when they're cajoled to satisfy their inborn affinity needs with modern Nature-abandoning habits. Their tropicmaking upbringing prevents them from relating non-destructively. It injures the Planet within and around them.

- As we discern the Planet as laws of Nature, we have in turn become subjects of these laws instead of fully free organisms.
- As we perceive the Planet as natural resources, we become resourceful exploiters of Nature and each other, instead of cooperative members of a supportive Earth community.

- As Nature becomes vulnerable to the objectivity of science, we lose the value of our feelings.
- As Nature succumbs to our power, we become subservient to those in power.
- As we destroy Nature's community, we lose our sense of community; as we alienate ourselves from Nature, we alienate ourselves from each other.
- As we capitalize on Nature, we become more capitalistic and less humane.
- As we compete for Nature's raw materials, we learn to relate competitively.
- As we treat Nature as grist for our civilization, we learn to treat each other uncivilly.
- As we identify ourselves as masters of the Planet, we lose our sense of place.
- As we act like a toxic, runaway cancer, eating the Planet's life away . . .

People's Inner Nature responds to being imprisoned as a slave servicing tropicmakers. It often feels attacked, angry, abandoned, worthless or incompetent. It cries for help and it gets tropicmaking medicine. That's like knowing carrots are good for your eyes, but it hurts like hell when you stick them in.

The individual must immediately respond to the pain. Often people pacify their Inner Nature's discomfort with whatever band-aids are available: aspirin, work or relationships lacking in meaning, drugs, food, alcohol, a streamlined physique, greed, excessive technologies, power, isolation, entertainment, clothes, money, status, violence, inferiority feelings, depression, escapism, excessive cleanliness, ad nauseum. **Often these band-aids are natural affinity relationship or love substitutes.** They become behavioral habits that artificially provide feelings of immediate worthiness, love and comfort. However, **most artificial pacifiers contain natural destructive elements. With the pacifier, the person inherits these destructive side effects as habitual personal problems.**

People seldom recognize that their personal or global problems are partially caused by their acceptance of modern society's tropic-making abandonment of Nature. Nor can they learn that lesson from their society. Prejudicially, modern society considers attacking or improving Nature to be an applaudable, non-controversial fact of life. It misguides our economic growth, scientific objectivity, standard of living, lifestyle, international relationships and our too appropriately named gross national product. It misguides us personally, too. That's why the cost of living is your income plus 20 percent.

Modern society attacks the natural systems upon which our health depends. The relationship is a global and personal Catch 22. But it's resolvable in the map's golden area. This area allows a person the time and space to revise their map and habits, to become aware of the whole situation and then choose to better relate to their Inner Nature's pain and needs.

The golden glow connects us not only to Nature's affinities, but also to Nature's powerful time, the NOW. It strengthens our Earth kinship by making it a thoughtful act. For example, picture yourself riding a surfboard atop a mid-ocean wave. During this NOW moment on the board, look back and see a log that you almost ran into. It floats far behind you. Ahead you see a fishing boat that you may reach in the hours to come. Dolphins play in the shining water beside you as you shift, thrust, balance and steer to adjust to the ocean's temperament. You are fully acting in the present and enjoying it.

Like the event of riding a wave, Nature within and without is the immediate moment of the ride, not before or after it. In Nature, time and action synchronize; Nature knows time as actions and relationships, not concepts. Time as we know it in modern society is nonexistent in Nature. For example, in Nature an ocean wave, a rock, a tree and time are exactly the same thing. The interrelationships of the millenia lead to and produce each of these entities in the present. The global interactions of wind, water and weather systems enact the NOW event of the wave. Because in the long run Nature is an ongoing **act**, it best understands **acts**, not well wishes, intentions, aesthetics or abstractions.

Being What We Feel, *Bowmansville, Pennsylvania*

Modern people usually know time as an abstraction, a concept, not an act, for memory and our state of consciousness create time by mentally extending the dimensions of Nature's immediacy. **Our mind imagery conceives of the past and future, but they are**

images, not realities. For example, when you ride a wave you can see the waves behind you, where you've been, like viewing the past. You can also see the waves before you, where you are going, like viewing the future. But Nature and your Nature is the present, the event of you riding the wave and seeing the past and future while riding it.

Your tropicmaking mentality may act by dwelling on the past or future, but only **acting in the present** connects you to the wholeness of the Planet. It connects your Inner Nature with Planet Earth's time. For example, acts of the past can't be changed. While riding the wave you can do nothing about the log behind you. And most often, the future turns out to be different than expected. Will the fishing boat be there when you get to where the boat is now? How often has your future turned out differently than what you had hoped or planned? How much energy did you expend in the planning? How much did you manipulate or exploit your Inner Nature or Planet Earth to meet your expectation?

Nature and the golden glow are the present, the wholeness and integrity of the ongoing wave. It can heal rifts between your upbringing and your Inner Nature, for only in the present can we perceive, think, feel, act and affect our situations; in the present, we are them. In the NOW of Nature we can consciously change our environment, select positive symbols and images from Nature, move tension-producing images out of our consciousness and disassociate from their signals. In the present we can begin to make a habit of this to the relief of our inner tensions and planetary tensions, as well.

The golden glow also evolves a new language that helps to bridge our culture's separation from Nature. Called **I choose to**, the language consists of using these three words silently or aloud. **I choose to** helps you and Nature grow, mutually producing a greater diversity of cooperative life-supporting relationships. It gives Nature the chance to be recognized. (You can give the language a mysterious, Eastern-type name, "ICHUZTO," if you choose to.)

I choose to symbolizes an aspect of our lives that applies personal freedom of choice to feelings, symbolizing and actions, and thereby to our society's relationship with the Planet. **I choose to** has two parts: 1) the spoken or deliberately conceptualized symbols "I choose to," and 2) the value, "I approve of me speaking, **I choose to**."

Although **I choose to** appears simple, its profound effects change your world because they can influence the NOW of Nature and its strained relationship with modern society. **I choose to** gives Nature a chance to grow by letting you choose to give Nature recognition for its contribution to your life. For example, if it's raining and you're getting wet, instead of being subject to discomfort from the rain (or from culturally tainted images of nasty, foul, harsh or bad weather) you instead rule your uncomfortable feelings by **choosing** to feel okay about rain, or you choose to get wet, or choose to sing, or choose images that offset your discomfort with rain, or choose to think about the values and fun of rain, and then you validate your choice.

In general, you can choose gratifying signals and associations (symbols, thoughts, people, places, memories, projections and things that evoke immediate or long-term gratification). You can choose to reject discomforting stimuli and situations. You can choose to enjoy Nature's fluctuations and objects. The act of choosing is your ultimate psychological security because it, in turn, lets your rationale support your Inner Nature's affinities. You make the present. You choose the words or events that will make and keep your Nature comfortable.

Repeated use of **I choose to** builds a new center within your psyche. It makes you whole by putting your sense-making abilities into action with respect to harmonious survival. **I choose to** guards you by placing your rational self between threatening aspects of modern life and your Inner Nature. That's rational, isn't it? It gives your Inner Nature confidence in your protecting it. In return, the Planet gives you its joy.

Psychotherapy deals with feelings, yet therapy, as it is commonly practiced, is often a band-aid, not a cure, when it doesn't deal with global integration of basic survival feelings. Although it definitely helps to make time and space for a healthy psyche and togetherness, it usually does not include the map. The map is predominantly nonexistent or taboo in psychological circles because, like everyone else, most psychologists are trained as tropicmakers. They're part of the problem. They inadvertently demean Nature because they are not trained to use the words "Nature" and "feelings" interchangeably. They don't recognize that what they treat is the discomfort caused by excessive tropicmaking, attacking and abandoning Nature in people and the Planet.

Therapy usually takes place in a red portion of the map. There it deals with red feelings. Often it trains people to adapt happily to tropicmaking's bizarre, artificial world. But becoming immune to its hardships or cruelty does not necessarily prevent them. Can a person really be happy when surrounded by excessive global stress and unhappiness? Can the misled lead the lost? Not without an accurate map that both can use.

Yet, you can make your life and the Planet's healthier. Share this book and its ideas with counselors and psychologists you know. Then you can counsel each other. Just don't expect payment for your efforts.

Community Spirit, *Tonto National Forest, Arizona*

Invoking the golden glow is essential when a person first becomes aware of any stressful feeling. It provides time and space to **locate that feeling's natural origins,** for therein lies any feeling's global source, survival value and regulatory feedback. The golden glow combines with an individual's blue natural feelings, becoming a green consciousness of these feelings. It's Gaian Social Security.

Green expression makes for a healthy person and a healtny Planet, too. It consists of thoughts, feelings and actions that harmonize. They connect your Nature within to Nature without. They allow you habitually to surround your Nature with supportive affinity images and relationships, while rejecting those that abandon Nature. Instead of causing modern stress and destruction of life, they produce a more lasting joy. You become a green Santa Claus. Try it and see for yourself. The next chapter explores how to evoke it.

Although the map's process is extremely rational, it's not completely scientific because for each person the golden glow is subjective, since it makes one conscious of one's personal and global history, habits and feelings. But if you need further evidence that your internal map is your destiny, permit yourself to try the following.

Hold your hands evenly straight out in front of you so that they parallel each other but do not touch. Ask your Inner Nature to keep your hands in that position throughout the exercise. Now close your eyes. For a minute, imagine placing some very large dictionaries, each weighing ten pounds or more, on your left hand, one after another. Keep the heavy dictionaries weighing on your left hand.

Now imagine tying helium balloons to your right hand—more and more helium balloons, thirty in all, pulling your right hand, making it lighter and lighter. Now add one more heavy dictionary to your left hand and four more balloons to your right hand.

Open your eyes and observe the position of your hands.

The above event often demonstrates how your conceptual self affects your Internal Nature without your even realizing it. Usually by the experiment's conclusion, you've lowered your left hand and raised your right hand, without ever realizing they had moved.

You can also experience your conceptual map's influence by using two identical objects of equal weight—two spoons, for example. After doing the dictionary and balloon imagery, lift one spoon in each hand. Often the spoons appear to have different weights. In the same way, the unrevised tropicmaking map causes our problems. It distorts our relationships with the global life system. It tells us to jump off a tall building to show the world we have guts.

STUDY GUIDE—CHAPTER 12

1) The golden glow suggests that your Inner Nature is compatible with natural systems. The exercises that follow allow you to use symbols and feelings to connect the two. They always work. If they don't seem right at first, you have probably located an area in yourself that needs your affirmation. Take time to find personal examples and convince yourself that your human nature is O.K. in areas that don't feel right.

2) Try to remember your emotional traumas, problems or self-putdowns that Nature healed with time. Are these traumas ever brought to consciousness by similar stressful situations? Do you still put yourself down?

3) Permit yourself to use Nature to help you respect yourself.

a) In Nature find examples of skills, beauty, survival, intelligence, dependability, creativity, power, friendliness, honesty, nurturing, physical skill, warmth, cooperation, love, loyalty, independence, freedom or other positive human aesthetics, values or aspirations.

b) Write a sentence explaining why or how some aspect of Nature is one of these values. For example, "Granite is dependable because it maintains itself during winter and summer."

c) Substitute your name for the natural object and permit aspects of your being to fit the description. For example, "I am dependable because I maintain myself during winter and summer."

d) Because Nature is in you and is always positive with respect to survival, as in Chapter 8's Planet-person reading, each of these sentences will work for you. When here, or in Chapter 8, you find areas that you can't comfortably label yourself as you labeled a natural entity, you have discovered an area of yourself presently lost to excessive tropicmaking. Work on regenerating these lost areas. They still exist in you; you just no longer see them. They strengthen your self-concept. Language is destiny. For further highly recommended exercises, read Cliff Knapp's **Creating Humane Climates Outdoors: A People Skills Primer***[11].*

4) At three different times during the day, take a few minutes to bring your consciousness as fully into that moment as is possible. Think of yourself and your surroundings as one giant ongoing unity. Write down what you are aware of at that moment. After a while, do you feel more in control?

5) Give yourself a gift.

    a) Give yourself a gift or gifts you appreciate from a natural setting (shell, flower, rock, etc.), and explain to yourself what there is about you that at that moment desires and therefore deserves the gift.

    b) Describe yourself to yourself as a person who finds joy in some aspects of Nature. Be sure to appreciate that aspect of yourself which values the gift(s) you have selected.

NOTES

On this page or elsewhere, write down any thoughts or feelings that have come to you from reading this and previous chapters.

Enough Thinking, Commence Relating, *Death Valley, California*

CHAPTER THIRTEEN
DANCING IN THE GREEN

Whether we like it or not, each of us learns and follows the symbols of a life map for survival. The increasing chaos of modern life suggests that our faith in the tropicmaking map is misguided.

Furthermore, maps are only part of making our way through the world. Like the Planet, we survive by doing it. We form relationships that allow life to happen. We become a statement to that effect. Relationships consist of thoughts, feelings and actions, not only maps. The best map is one that makes one think: "enough intellectuality, commence relating."

Our revised consciousness map depicts that the Planet finds its way to stability by using affinity relationships. The map's potential is that it lets affinity speak by making room for us to comfortably or uncomfortably sense it. Yet we must recognize that like learning any new language, learning to speak affinity takes time, diligence and motivation. For this reason, to learn the ways and wisdom of the natural world, the blue portion of the map, one must spend time enjoying habit-forming relationships with it.

Personal security is found in recognizing that you embody the Planet. It's knowing how to let your Inner Planet safely enjoy natural affinity relationships and how to select or create environments that allow them to take place. It permits you to find and enjoy the natural world that exists within yourself, in every other person and in the natural environment.

With respect to yourself: Although you can't easily visit your liver, kidneys or spleen, you can get to know their cooperative affinity through your heart. It is the wilderness within. Its pulse conveys the T-R joys of life.

With respect to other people: Friendships between people are natural affinity feelings. Sticky, stressful interpersonal relationships usually occur between the blue and red portions of people. When possible, disregard the nonsensical red rules and customs of tropicmaking. They are the red menace whose song is, "increase your dominance over, and separation from, the natural world." Be careful as you sing that song.

Go back to Chapter 2, and see if the map now allows you to discover the abandonment of Nature which is causing the list of problems people experience. Remember that people's Inner Nature and the Planet are one. The Planet endures the same problems and hurts as do people. It desires to eliminate these problems just as much as we do. To that end, the map integrates our efforts. Through it, we discover how Nature heals its wounds and ours as we learn to share Gaia's wisdom and life experience while gaining unity, peace and common ground.

Relate to others by sharing the map and knowing where you are on it when things get sticky. Usually one or both partners are acting out or imagining RED and reacting accordingly. Let your Nature speak directly to the Nature in others. That relationship is an ongoing love affair. Blue and green resolutions can be mapped out.

People can relieve red relationship's stress by seeking the Nature they share in common. Relationships with people who won't seek or enjoy their blueness are questionable. The time may be better spent enjoying green relationships. Revere the blue and green aspects of your life; question or confront the red at home, work or school. Whenever captured by redness and powerless to escape it, teach yourself to enjoy confronting and contesting it while coasting with it. Make it a fun game and learn to laugh. Get some good out of everything. If life hands you a lemon, make lemonade, or furniture polish. Ask yourself for permission to take risks; don't let risk taking capture you by becoming a habitual outlook, rather than a choice.

With respect to Nature: Participate in the natural world. Let your Nature know and enjoy its Planet Mother. Don't fear Nature just because you can't always control it. Every one of your positive thoughts and feelings about Nature represents the real facts of life. Mother Earth is a planet, not a parent. She won't nag you to wear clean underwear in case you get hit by a car.

Survivalwise, Nature can do no wrong, in the long run. Give the Planet organism room to express its T-R relationships as it has over the millennia. It and the stars are the essence of you. Let them shine.

Natural habitats and areas are organs of the Planet. Each shares its peace and tells its story. Nourish your good feelings and make more connections with Gaia's global network of life sensations. Let natural areas and their "long waves" touch you in every possible way. Their biological science and identification are only part of the picture. Their smell, taste, feel, color, motion, uniqueness, life, beauty, consensus, power, integrity, ethics, honesty, relationships, affinity,

Look Out to See In, *Golden Canyon, California*

shape, form, design, temperature, changes, fun, adventure, wildlife and existence are all a part of you that's hard to find without them. That's why they're inviting . . . and essential. Enjoy watching the fly fly; become the spider spinning its web.

Create exotic nature trails and outings that give people access to the information in Chapter 8.

Every natural setting is the affinity expression of five or more billion years. For the same reason that you are alive from the sexual love affinity expression between your human parents, each natural setting also lives from past affinity relationships. Your positive sensations about Nature are the Planet's life expressing itself in you. Your feeling about the clouds, sky and stars exist because they exist. Your ethics and morality express your desire to live harmoniously with the Nature in people. They must also include the Planet's life.

When you enjoy natural areas, your good feelings, thoughts and actions become part of that setting's affinity. Liking, loving or revering Nature's ways expresses your affinity for life. Allowing yourself to fully feel good about natural areas helps these areas survive. We don't like to hurt or kill what we or others love.

Tropicmaking is a subversive activity. It denies Americans the freedom to relate to Planet Earth in a mutually beneficial way. Instead, it cajoles us into destroying the underlying source of our survival and joy: Nature—**the art of God**. Each year in America, we pave five thousand square miles of countryside, while globally, seventeen thousand species become extinct. We plunder, not harvest, irreplaceable natural areas at will. Just as excellent Army basic training makes us excellent soldiers, our excellence in education often makes us excellent tropicmakers. We let it happen because we don't recognize that modern society's sickness has us fighting its undeclared war with Nature.

Each natural entity is a feelingful part of yourself that you lose when it's gone. When your life feels lonely, drab, frightening, stressful, isolated, depressing, hurtful, hateful or generally negative, it's always because your tropicmaking habits and relationships have overwhelmed you. They have abandoned your Nature, just as they've abandoned natural areas. Excessive tropicmaking is destructive, but it's fair. It subdues all forms of Nature, including your good feelings about being alive. It makes us neglect Nature like we neglect the subtitles on erotic, foreign films.

Your best long-term defense against feeling bad is regenerating stabilizing affinity relationships with Nature, your biological home, the fountainhead of your positive feelings and sensations. Returning to our natural origins for peace is its own reward, and it's not new. Jesus spent most of his time in Nature, alternating between teaching the people in their towns and retreating to pray and meditate in the hills.[1] But it wasn't retreating, it was regenerating kinship with the natural world. And what was true for Jesus is true for most other spiritual leaders, as well.

Moses received the ten commandments in the mountains, not downtown. John the Baptist was a preacher from the Judean wilderness: "A voice crying aloud in the wilderness," who came to "prepare a way for the Lord; clear a straight path for Him. John's clothing was a rough coat of camel's hair, with a leather belt 'round his waist and his food was locusts and wild honey"(Matt. 3:1-5).[1]

After Jesus was baptized by John the Baptist, he went out into the wilderness for forty days and nights to fast and to triumph over the devil's temptations. He emerged strengthened spiritually, prepared to live out God's promise.[1] Strengthened spirituality with Nature is just another way of saying regenerated kinship.

Jesus said, "Put away anxious thoughts about food and drink to keep you alive, and clothes to cover your body . . . Look at the birds of the air; they do not sow and reap and store in barns, yet your heavenly Father feeds them . . . And why be anxious about clothes? Consider how the lilies grow in the field; they do not work, they do not spin, and yet, I tell you, even Solomon in all his splendors was not clothed like one of them. But if that is how God clothes the grass in the field, which is there today, and tomorrow is thrown in the stove, will he not all the more clothe you?" (Matt. 6:25-30)[1]

Nor is the Old Testament to be ignored. "Or speak to the Earth and it shall teach thee," we can read in Job 12:7.

"Thy wrath is come . . . and shouldst destroy them that destroy the Earth." (Rev. 11:18) Recognize that Nature has biologically designed us for harmonious survival with the Earth organism, and if we can't rationally actualize our inborn Planet-person harmony, then for self-preservation, Gaia will modify our activities through its actions which we call natural catastrophes. Don't wait until you're dead to be part of the Planet through green relationships. Learn to make it your home while you live. Let yourself feel what you see happening to Nature. Your biological home inside and outside is under siege. Relate from these feelings, not only from your artificial ones. Let the life of St. Francis of Assisi set an example for you to follow.

Our right-handed thinking is underhanded with respect to our biblical interpretations. It teaches us to revere the Bible's genesis of Eden's garden and Adam and Eve. In Genesis 1:27 God says the way to live is to "Subdue the Earth . . . and take dominion over it." Be aware that too often, Madison Avenue is tropicmaking's CIA.

Tropicmaking's bias ignores that with the great flood (Gen. 7:23), the story changed. God completely erased the original world He created because He made a mistake. He gave people the wrong orders, so it didn't work well. How could His Garden of Eden grow, if He told people to subdue it and take dominion over it? After all, it's hard enough to maintain a garden when the gardener is trying to nurture it.

The Bible tells us that with Noah and the great flood, God remade the world. To Noah's family, God said, "Bring forth with thee every living thing . . . that they may breed abundantly in the Earth and be fruitful and multiply upon the Earth." (Gen. 8:17) "And replenish the Earth." (Gen. 9:1)

Get out your Bible. Read it for yourself. We are basically the children of Noah's world and family, not of Adam and Eve's. The flood removed the latter and their order to dominate Nature.

Read further and you'll discover that God never commanded Noah or us to subdue the Earth, nor to take dominion over it. To

Muddy Water's Challenge, *Congoree River, South Carolina*

us and the new world He made for us, God said, "Conserve the Earth and replenish it." And that's what Noah did. It worked, too. Many generations later, "The whole Earth was of one language and one speech." (Gen. 11:1)

Perhaps modern tropicmakers interpret the story their way in order to justify our excessively exploitative relationships with the Planet and each other. But as human finalists in life's race, let's be the nation we really are under God. Let's be Noah's children and elect a President and Congress dedicated to spending the national budget on replenishing people and the Earth from excessive tropic-making. Today we and the Earth suffer because too often we elect to office our most capable tropicmakers, not our most dedicated environmental leaders. Let's elect officials who know how to deal with our destructive habits. There's about 150 years worth of wonderful jobs, goodwill, prosperity and peace in that policy. They give us a green light and a golden future using state of the Ark technologies. It's a workable response to militant communism and violence. The latter only attract unhappy people whose life affinities have been unfulfilled or exploited by tropicmaking. Their way doesn't bring them greater happiness.

With respect to human impact upon Gaia, it is important to note that scientists report that nations with intact ecosystems pay less to produce their gross national products because of their attitude and Nature's ability to recycle waste products free of charge. Industries that show active concern for the environment are usually good investments because their management acts responsibly in all areas of endeavor, including economics.

Like a blade of grass pushing its way through an asphalt road, the delights of natural relationships can guide our survival. When they become disharmonious and confrontational, survival tensions appear, signaling danger and preventing destruction. For example:

• Overpopulation tensions have been shown to make deer, bears and people sterile until their population's size can be accommodated survivalwise.

• During the Second World War, the threat of Adolf Hitler's Nazis made us do what today seems impossible: to nationally grow personal vegetable gardens and be friends with the Russians.

• Chernobyl and Harrisburg dramatically made many of us concerned about nuclear safety and international peace.

• AIDS is changing our sexual habits and tending to stabilize relationships, as once did the family farm.

• In our society, money is survival. When the government placed withholding taxes on our bank accounts, 22 million Americans arose as one and wrote their congresspeople. They repealed the law immediately. That wasn't scientific, but it is a fact.

These examples demonstrate active survival feelings' latent power. They govern all other species and the Planet. They're similar to the survival feelings of a priest who lived in a tough neighborhood. He said, "My crucifix protects me from muggers because I keep it moving at a dead run."

Never have any taxes, laws, wars or developments taken place when enough people felt they didn't want them. We only live in an outrageous world because our upbringing has destroyed our natural power to feel outrage about Nature's destruction.

Our natural survival affinity feelings have been lulled into becoming tropicmaking's artificial feelings—obedience and apathy. They have been bought. Our love for Nature has become "I love to shop, I love new things, I love money, I love power, I love clothes." But love Nature and it's like having an illicit affair. In fact, we usually have more survival feelings attached to a dollar bill than to the global life system.

Tropicmaking teaches us to live almost entirely in the world of symbols and images. The 3 R's are required because we think that they are survival. For example, a person with a Ph.D. in Native American studies may intellectually know about what an Indian medicine man can do. The Ph.D. holder gets a prestigious college

Consensus Embodies Earth Processes, *Canyonlands National Park, Utah*

job, while a real Indian medicine man may be on skid row because he feels useless to a tropicmaker's society.

It has reached the point where today some of us believe ourselves to be acceptable if we look like the images that appear in the media. Today's fashion for men is to have an unshaven look because some character in a television series looks that way. An electric razor is on the market that gives you the look of three days' unshaven growth. This razor has made headline news in the media, along with another new invention: an electrically heated toilet seat that maintains its temperature at 80 degrees. The seat allows you to exercise your freedom of choice, at least with regard to temperature! You may adjust the toilet seat thermostat higher or lower. Isn't that wonderful?! Next thing you know, genetic engineering will cross pasta with a boa constrictor and get spaghetti that winds itself around the fork.

One of our biggest problems is the industrial trash dumped into our rivers, stores and minds. The natural world and the Nature within us is being devoured for the production of these kinds of images and products. Without active, constructively expressed outrage, how will we ever draw the line?

Technologies Can Relate, *Lubec, Maine*

Yet even though we know the irrationality of destroying our life system, our upbringing refutes our discomforting affinity feelings' value. It labels them childish, sentimental silliness or feminine. Prove them scientifically, mathematically or economically—otherwise, express them to your pet, garden or Teddy Bear. Mature tropicmakers don't take them seriously, and for God sakes, don't express them at an environmental impact hearing.

One who knows the Creation knows the Creator. God is an intense affinity emotion that creates global life as we know it. But, without your feelingful resistance, tropicmaking uncontrollably disrupts life. **It would like you to forget that survival is an emotion.** In too many ways, our lives are cannon fodder for the economy and conquest of Nature. However, do not be alarmed, for excessive tropicmakers have a wonderful label for their activities. They call it "PROGRESS." And anything that further separates us from the natural world, no matter what its negative impact, they call "progress."

They bring to mind the story of a horse who used to walk into walls, trees and fences. Its new owner returned the horse saying, "This horse is blind."

"No sir," replied the seller, "that horse ain't one bit blind. It just don't give a damn." Is that horse's name, PROGRESS? For if it is, we must recognize that progress which loses sight of our kinship with the Planet is a horse of a different color.

If our internal organs disregarded each other's signals, we would die. The T-R affinity that makes organ systems a cooperative community also organizes, maintains and regenerates global life. People know this affinity as their comforting or discomforting feelings. Thus, learning to feelingfully appreciate the natural world, and live via these feelings, is urgent for global peace and survival. It allows us contact to resonate with the natural world. We must protect and defend intact natural systems with constant pressure, constantly applied. We can prevent our environmental jams with wildlife preserves.

Recently, I heard about a large tree whose washed-out roots were in the shape of a human body, for a person had been buried under it. The body had nurtured the tree whose roots, fully infiltrating the corpse, assumed its shape. Many people gave special feelingful respect to the tree, recognizing that it and the buried person were one. They protected the tree, treating it humanely, as if it were sacred. This suggests that if we initiate practices like burying our loved ones in natural areas, we can invoke new humane feelings and unity with these areas. We need to develop modern rituals that detach our affinites from the greed of excessive tropicmaking and reattach them to the Planet's integrity. Such rituals modify our exploitative desires because they renew our self-regulating spiritual identity with the land and with each other. Through them the landscape again becomes a cathedral which we can understand and which thereby speaks to us in its ancient protective ways. Ritualistic bonding with Earth regenerates the kinship that prevails in other, more environmentally and socially sound societies. [6:Johnson]

Out of dedication and delight, some people live fulfilling, nurturing lives directing nature centers and sanctuaries. These diverse green areas give so much to so many species, yet sometimes we totally eradicate them to introduce a new single species, **Interstatei cloverleafus.** It produces a miraculous habitat that is constructed by caterpillars and attracts many unexpected forms of wildlife: pintos, cougars, mustangs, colts, firebirds, impalas, eagles, foxes, bugs, beetles, lynx, rabbits, falcons, rams, phoenix, broncos, thunderbirds and even barracudas.

In our own way, we must each become our own naturalists. A favorite story conveys why.

The World's Smartest Man finds himself together with the President of the United States, a priest and a naturalist in a plane doomed to crash. On board are only three parachutes, which prompts the President to take one immediately and jump, with the declaration that he owes it to the American people to survive. The World's Smartest Man steps forth next, claims that his life is an irreplaceable asset to humanity, and exits. The priest looks at the naturalist and says, "I have lived my life and now it is in God's hands; you take the last parachute."

The naturalist replies, "No sweat, Father, we're both safe. The World's Smartest Man just jumped out wearing my backpack." The state of the environment makes this story believable.

Habits resist change. Because our programming makes us think and act in habitual ways, our perceptions of the Earth and our relationships to it change slowly. **But we ask for more trouble if, educationally, we continue to mislead our children about the nature of themselves and the Living Planet.** It is morally wrong to mislead young people about the possibilities of their lives. Today, wisdom means not compromising the natural integrity of organism Earth within and around us.

For peace, health and survival, our education must include active, experiential, holistic involvement with Nature in our backyards and globally. It could emulate the National Audubon Society Expedition Institute, a demonstration model that achieves these goals as it evokes the golden glow. By encouraging our students to validate and share their behaviors, feelings and secrets, the program permits its participants to re-discover their Inner Nature and connect it with the Planet's life.

At the Institute, as we recognize society's negative impacts, we help each other design custom-made means of dealing with our personal and collective problems. To accomplish this, we allow our Inner Nature to express itself without harm. Otherwise, it remains hidden and unreadable. By contrast, openness and honesty make it possible for us to choose habit-forming green relationships and behaviors that:

- recognize feelings are facts of life
- recognize feelings are rooted in the Earth
- recognize Planet Earth as a self-organized living organism which inhabits us biologically and emotionally
- validate our natural sensations, feelings and reverence for life; let us substitute the word Nature for feelings
- help us recognize that affinity feelings about our natural selves or Nature are global survival feelings; recognize our negative or stress feelings as our Inner Planet's reaction to being abandoned

- satisfy our natural feelings without excessively using technologies; give us confidence in the Planet's self-organized, sentient survival wisdom
- engage in technologies and lifestyles which give joy while harmonizing with Nature, and therefore produce little garbage or waste
- satisfy our natural survival feelings rather than our artificial acculturated feelings or womblike tropical memories
- are based upon choice rather than habit or conditioning
- recognize both our human and Earth mothers
- celebrate the natural world within and around us and in others
- recognize affinity, cooperation and peace as functions of Nature
- feel comfortable as these relationships are measured by their long-term effects on Nature and humanity and their ability to support life
- recognize and celebrate Nature's role in every act

To Replenish, Not Subdue, *Bradenton, Florida*

- environmentally re-educate ourselves by validating Nature's timeless life wisdom
- confront and subdue cultural fears and prejudices against Nature within and without
- disengage us from stressful past and future messages or expectations by focusing on the NOW of life
- disregard media messages that demean or victimize Nature within or without
- make life and the landscape sacred
- teach us to commune with Nature within and without on all levels and in all activities
- give us security based upon our ability to laugh at modern society's shortcomings
- do research that supports, not compromises, Nature
- create good feelings about ourselves as natural beings

Kinship Through Stewardship, *Bowmansville, Pennsylvania*

- strengthen our natural affinity attributes and our culture's survival responsibilities
- govern by consensus, which includes each person's thoughts and feelings
- establish counseling or stress-resolution settings that identify our Nature, feelings, emotions and experiences with those of the Living Planet
- establish Nature-congruent symbols, communities, interpersonal relationships and wilderness experiences
- counteract our civilization's desire to use excessively the remaining natural world as grist for converting the temperate and arctic zones into "tropic-like" environments
- make physical and psychological space to reconnect our Nature within to the wilderness community in people and the Planet
- help us recognize our society's effects upon ourselves by measuring its effect upon the natural world
- celebrate Nature-enhancing cultural differences and diversity
- provide equal time for Nature at home and in school
- recognize that consciousness is a Planet function
- recognize NOW as a blend of Nature, culture and survival affinity
- get involved with environmentally responsible political, scientific and social processes
- build our own educational roads by using Chapter 11's map—transfer our modern survival feelings away from tropicmaking and reunite them with Gaia

Note that on the expeditions, these directions are not idealized goals. Rather, they are **actualized, nurturing relationships, occurring because the participants build trust in their feelings, Nature and each other.**

Students graduating from programs like this enter all fields and walks of life. They succeed by fully developing and taking advantage of the environmentally sound, but sometimes less financially rewarding, aspects of their professions and communities. Over 60 percent of them become natural scientists or outdoor educators. Many act politically and join movements for a healthier world. They persevere because their feelings for life motivate them. Excessive tropicmakers usually disappear when they lose money. Education is a more permanent solution for environmental and social stress than is legislation, for the latter is too often blown away by political winds.

I propose that each of us who is under stress of any sort create or join an Earth Kinship Affinity Group (EKAG). These groups of ten to fifteen persons operate by involving their members in the educational processes listed above. EKAG groups meet once or twice a week as affinity and support situations for their members and for the Planet. Using our Planet-person map, group members help each other identify the aspects of their daily lives that stress their Nature. They also identify how these stress factors hurt the Planet. Without guilt or fear, they acknowledge how they feel. Then members *act* to relieve the stress on *both* themselves and the natural world. EKAG groups can exist within any type of organization or community setting. Guides trained to catalyze and facilitate them are available from the Audubon Institute and other whole-life organizations. EKAG groups help their members to constructively express the aspects of themselves that are Planet blue. They help regenerate kinship with the Planet. [see Appendix B]

Our love for the natural world is inherent. To discover and express this affair of the heart is one of life's greatest pleasures. Endlessly repeatable, it's why most kids, cultures and species want contact with the out-of-doors. Our devotion is neither childish nor uncivilized. It is life. We respect and cherish life by exercising it. We don't need a master plan of how to do this. When an uncomfortable entity appears that threatens our children, family, bank account, pet or car, we know exactly what to do, and we do it. The same holds true for our global Living Earth home and family, once we revere their existence. And, if what you want to do for Nature feels like civil disobedience or a small sin, do it. Remember, if you want forgiveness for your sins, you must sin first.

Earth Firster, Mike Roselle, who cannot count the number of times he's gone to jail for protecting forests, claims that going to jail for civil disobedience is like sex. "If you do it enough you can't remember how many times you've done it, and if you can remember how many times you've done it, you haven't done it enough."

Reaffirm your kinship with the Planet. Validate and treasure the good feelings you gain from Nature's beauty, motion, excitement, strength, power, freedom, adventure, age and wisdom. They are you. To deny them is a form of insanity, a national psychosis that converts Nature into Barbie dolls.

The planet organism called Earth best knows and understands diverse natural areas. Excessively, artificial places usually destroy or poison it. As we technologically attempt to save the Earth from its allegedly doomed relationship with the dying sun, we live out a destructive, self-fulfilling prophecy. For excessive tropicmaking, not the sun, is the immediate danger to Gaia. It is changing Gaia's purity, temperature, acidity and genetic pool with dire consequences forthcoming, such as flooding from sea-level changes, glaciation, desertification, weather extremes, toxic air and water pollution, and drought. These catastrophes but intensify the problems we already face. They are already taking place. The only economically feasible devices we have to reduce tropicmaking's excessive carbon dioxide are trees; and we're cutting, slashing and burning them, making more carbon dioxide, not less. [see Appendix C]

If we continue to act like a detrimental infestation, Gaia, the planet organism, as would any organism, predictably will prevent us from further harming it. That is the self-protection that typifies life. We must protect and fertilize forests, using rock dust, not further destroy them and ourselves. We cannot afford to wait for catastrophes like desertification in Africa or topsoil loss to trigger our survival feelings, for by then it's too late to rectify the situation. For survival,

Earth wants and needs the temperate zone to be temperate, not tropical, whether modern society wants it that way or not!

Consider this scenario; is it fact or fiction? Throughout human history, tropicmaker's separated-from-Nature artificial environments have excessively destroyed the natural safeguards and relationships that prevented the global spread of human epidemics like smallpox, plague and typhoid. These diseases were not usually found in aboriginal populations because of the latter's isolated, uncrowded, natural life ways and immunities. Artificially, we innoculate ourselves to build up our antibodies against microorganism epidemics. Thus we resist the efforts of microorganisms that might otherwise destroy our globally stressful modern societies. Not surprisingly, the global organism fights back against our efforts. After all, to Gaia we are often an uncontrollable, excessively tropicmaking infection that is giving Gaia a high temperature. Our modern ways often threaten Gaia's metabolism of intact ecosystems, clean air and pure water. Gaia fights us by unleashing a protective mechanism from the wilderness: AIDS. This virus terminates our ability to build up immunity to the Planet's natural safeguards, things which we call "epidemic diseases." In this way, through AIDS, Gaia prevents tropicmakers from destroying its global physiology. Isn't that exactly what our personal body attempts to do when we have a critical infection?

I don't trust tropicmaking because its short-circuited, basic axioms are "Planet Earth is dead until proven alive," "we may excrete in our own water supply in order to survive," "the world consists of mechanistic facts, not feelings" and "Nature is guilty of being uncivilized until proven innocent." That type of insanity underlies our insane world. We must purify the environment by cleaning up our act.[19]

Peace is built on feelings like trust. We trust because it is a personal sensation and experience. Unifying experiences build the security of trust. Entities that hold survival in common have the unifying grounds to trust each other. Our Planet is those grounds. We must trust its life ways.

A healthy Earth could continue its life for millions of years. I've heard no evidence to the contrary, and the Earth will tell you so, once you learn its affinity language. That's why in the long run, the Earth welcomes particular places or peoples that in the present support the global life system. We should show no compromise in defending Nature within or without from excessive tropicmaking.

We have yet to learn how to live with Nature. When our space program flies the rich folks off to settle other planets, they will carry our ignorance of Nature with them. They will no doubt fail there as we're failing here.

The true peaceful value of our green thoughts, feelings, actions, aesthetics and prayers is proven **when they actualize healthy natural land, air and water areas, not before.** If they don't vitalize and regenerate natural areas, they are misleading pacifiers that are part of the problem. Let's allow our lawns to grow wild! Let's make our lawn habits become habitats, and may our red habits turn green out of envy for life's blessings.

The battle that is causing global destruction originates in each person's mind map. It can be won in the golden arena by recognizing that Earth is a living organism which we embody, transferring our survival feelings back to it and then acting accordingly. We must treasure green relationships and places. They are survival. Our lives should be their statement, rather than that of destructive tropicmaking. Let us become as one with Earth and enjoy its pulse.

After all, if we don't learn to live with the natural world and the Earth gives out, where can we go? The moon? The moon looks like we've been there already.

STUDY GUIDE—CHAPTER 13

1) Use this chapter's list of National Audubon Society Expedition Institute goals as a personal checklist. For each goal:
   a) locate where you have had or observed that experience.
   b) design and make that experience happen for yourself now.

2) Speak with clergy or friends to further consider the author's contention that our world and people are Noah's children, not Adam and Eve's.

NOTES

On this page or elsewhere, write down any thoughts or feelings that have come to you from reading this and previous chapters. Then go back through your notes and

1) See how applying the map to them alters your impression of them.

2) Ask yourself how participating on a Nature expedition might alter your impression of them.

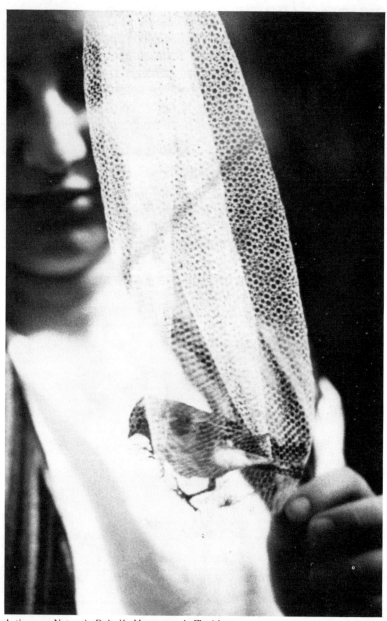

Acting on Nature's Behalf, *Homestead, Florida*

EPILOGUE

One means by which you can help your own integrity and Earth's is to photocopy the following petition, sign it yourself, collect signatures for it and mail it to: World Peace University, Box 188, Sweet Home, Oregon 97386. There, the names will be added to a master copy for submission to world leaders in 1990. I also urge you to embody the petition's message in a letter to your political representatives and other influential community leaders.

PETITION

Whereas undeniable evidence from scientific, historic, cultural and spiritual sources show Planet Earth to be a living organism (Gaia) whose global life pulse, physiology, psyche and health are personally embodied and enjoyed by all human beings, we submit that Organism Earth is:

- the only being of its kind in our solar system
- the active ongoing mother of life as we know it, and
- a principle source of humanity's health, happiness and survival.

We recognize that Earth is at risk due to excessively stressful human activities. It urgently requires the special ethical, moral and humane treatment accorded a benevolent individual or an endangered species.

By our signatures, we request that the United Nations and all societies honor Earth's life by:

- officially acknowledging Planet Earth as a living organism which organizes, perpetuates and regenerates itself,
- developing educational programs which teach people of Earth's life, its endangerment and how to reunite one's survival needs and feelings with Earth, and
- enacting strict laws which protect and preserve the global organism's integrity, for Earth's life and peace is our own.

Name   Address

_____   _____
_____   _____
_____   _____
_____   _____
_____   _____

APPENDIX A

**Information Concerning
the National Audubon Society
Expedition Institute**

If you have enjoyed **How Nature Works: Regenerating Kinship With Planet Earth,** the National Audubon Society Expedition Institute (NASEI) invites you to participate in its programs or help support them through donations. NASEI is a non-profit, approved and accredited educational organization, functioning cooperatively with Lesley College Graduate School in Cambridge, Massachusetts.

Wild America can be your campus for high school, undergraduate or graduate degrees and informative summer camping trips.

A one-room schoolhouse on wheels, the Audubon Expedition Institute is a small group of inquiring high school, college and graduate students and staff members who travel across the country as a learning community. As they camp out September through May and summers, hiking, canoeing and skiing, the Expedition visits peoples and places in order to observe, discover, question and develop a practical awareness of deep relationships within and between America's ecosystems, cultures and history.

Our Classroom is Wild America. You can't fully learn about the environment or environmental careers by sitting in a classroom. That's why the Expedition uses a hands-on approach. We'll teach you practical skills in ecology, community development, conservation, human relationships, energy use, nutrition, as well as the academics of the natural and social sciences, music and art, education and personal psychology. Seventy-five accredited courses are offered.

Who should apply? High school, college and graduate students or teachers who want to experience an environmental education alternative, who want to lead an ecologically sound life or who want to ask and find answers to important questions about the world in which they live.

The Expedition Education advantage incorporates all previous undergraduate course work, offers accredited career internships, independent study and life experience, and awards financial aid, postgraduate grants, A.A., B.S. and M.S. degrees. Courses emanate from real-life encounters, community process and environmental issues that involve coastlines, homesteaders, deserts, ski touring, appropriate technologies, wild rivers, observatories, research, academics, solos, lobbying, backpacking, traditional music, politicians, journal writing, contra dancing, national forests and parks, wilderness, Amish farms, mountains, Indians, coastlines and lots more. Students are admitted regardless of their race, sex, religion or national origin.

For a complete information packet, scholarship fund contribution material, information on one- to five-day workshops, summer programs and a student newsletter, write to the National Audubon Society Expedition Institute, Dept. Q, Northeast Audubon Center, Sharon, CT 06069. (203) 364-0522.

APPENDIX B

Information Concerning the Conference on Regenerating Kinship with Planet Earth

Announcing a yearly educational conference on a most vital issue of our time:

Regenerating Kinship with Planet Earth
on the beautiful Ciudad Colon, Costa Rica campus of the University of Peace, United Nations, with pre- and post-conference tours to natural wonders of Costa Rica and Central America.

CONFERENCE PURPOSE: To investigate, broaden and effectuate regeneration of people's kinship with Earth. The interdisciplinary, cross-cultural conference affirms, explores and applies the findings of Dr. Mike Cohen's book **How Nature Works: Regenerating Kinship With Planet Earth.** Dr. Cohen will personally conduct many of the sessions as well as consult with individuals and informal gatherings.

CONFERENCE PARTICIPANTS: All conference participants are expected to be proficient in the English language and to share their questions, thoughts, feelings and expertise when appropriate. Participants are invited to submit typewritten copies of their contributions to the conference upon its completion.

CALL FOR PRESENTERS: The conference seeks selected presentations on the subject areas found in **How Nature Works**. Presenters' applications should state the candidate's subject area, credentials, experience in their subject, past presentations at conferences, professional affiliations (if any), previous publications and a one-page abstract of their proposal. Presenters must signify their willingness to provide a four-page typewritten copy of their presentation, as well as permission for its editing and inclusion in the conference proceedings if they are published.

For further information write: WORLD PEACE UNIVERSITY, P.O. Box 188, Sweet Home, Oregon 97386 USA.

APPENDIX C

Information Concerning Ownership and Protection of Rain Forests

If you want to help protect rain forests by owning part of one, read further.

Rain forests play a vital role in the global life system by acting as a purifier of air. By removing excess carbon dioxide from the atmosphere, they help cool the Earth by reducing the "greenhouse effect." In addition, our Planet's beautiful equatorial rain forest greenbelt provides habitats for thousands of unique plants and animals, produces atmospheric oxygen and hosts most of the migratory species of birds during their flights. Without the rain forest's offerings, species disappear and, to our cost, atmospheric conditions change.

Because they are severely threatened by excessive deforestation, rain forests need your help. **One way you can help is for you to inexpensively buy and protect one of more of the rain forest trees.**

Several organizations now make it possible for you to buy, for only two dollars each, one or more deeds for a 5 x 5 foot plot of virgin rain forest land upon which grows a tropical tree. Your deed does not convey to you the rights to cut or harm your tree or your land. These rights are instead held by a non-profit wilderness protection corporation whose charter explicitly forbids it to cut timber or develop land. Thus, your purchase of rain forest property makes possible the most permanent purchase, protection and preservation of rain forest land. It makes a profit possible for the land's original owner without harming the land or the trees. In addition, it provides funds for the protecting corporation whose charter mandates that it use all collected monies, minus expenses, to purchase and preserve additional rain forest lands.

Your deed gives you visitation rights to your tree areas as well as a choice of the species of tree you desire to buy. It is suitable for framing and makes a unique, educational gift. It helps draw attention to both the contribution and plight of the rain forests and acts as an example for others to follow.

For further information, send a self-addressed, stamped envelope to WORLD PEACE UNIVERSITY, Box 188, Sweet Home, Oregon 97386 USA.

BIBLIOGRAPHY

1. Caduto, Michael. "The Bible's Mandate: Care for the Earth God Made." Lebanon, New Hampshire: Valley News, June 8, 1984.

2. Cohen, Michael J. Across the Running Tide. Freeport, Maine: Cobblesmith, 1979.

3. Cohen, Michael J. As If Nature Mattered: A Naturalist Guided Tour Through The Hidden Valleys of Your Mind. Sharon, Connecticut: National Audubon Society Expedition Institute, 1985.

4. Cohen, Michael J. Our Classroom is Wild America. Freeport, Maine: Cobblesmith, 1974.

5. Cohen, Michael J. Prejudice Against Nature: A Guidebook for the Liberation of Self and Planet. Freeport, Maine: Cobblesmith, 1983.

6. Cohen, Michael J., Editor. "Proceedings of the Conference IS THE EARTH A LIVING ORGANISM?" Sharon, Connecticut: National Audubon Society Expedition Institute, 1986.

7. Dr. Seuss. The Lorax. New York, New York: Random House, 1971.

8. Dyer, Wayne. Your Erroneous Zones. New York, New York: Avon Books, 1976.

9. Farb, Peter. Man's Rise to Civilization. New York, New York: Dutton, 1968.

10. Joseph, Lawrence. "Britain's Living Earth Guru." New York, New York: New York Times Sunday Magazine, November 23, 1986.

11. Knapp, Clifford. Creating Humane Climates Outdoors: A People Skills Primer. Box 313, Oregon, Illinois 61061, 1985.

12. Kroeber, Theodora. Ishi. Berkeley, California: University of California Press, 1961.

13. Lobell, John. The Little Green Book. Boulder, Colorado: Shambhalla Press, 1981.

14. Lovelock, James. Gaia: A New Look at Life on Earth. London, England: Oxford University Press, 1979.

15. Lovelock, James. Gaia: The World as a Living Organism. New Scientist, December 18, 1986.

16. Margulis, Lynn. Microcosmos: Four Billion Years of Microbial Evolution. New York, New York: Summit Books, 1986.

17. Murchie, Guy. "Secret Life of Rocks." Dublin, NH: Old Farmer's Almanac, 1987.

18. Murchie, Guy. Seven Mysteries of Life. Boston, Massachusetts: Houghton Mifflin, 1978.

19. National Wildlife Federation. Conservation Yearbook. Washington, DC, 1987.

20. Pearce, Joseph Chilton. The Magical Child. New York, New York: Bantam New Age Books, 1980.

21. Russell, Peter. The Global Brain: Speculations on the Evolutionary Leap to Planetary Consciousness. Los Angeles, California: J.P. Tarcher, 1983.

22. Samuels, Michael. Well Body, Well Earth. San Francisco, California: Sierra Publications, 1983.

23. Simmons, Leo, Editor. Sun Chief. New Haven, Connecticut: Yale University Press, 1963.

24. Van Matre, Steve. The Earth Speaks. Warrenville, Illinois: Institute for Earth Education, 1983.

25. Vernon, M.P. The Psychology of Perception. Gretna, Louisiana: Pelican, 1971.

26. Young, Louise. The Unfinished Universe. New York, New York: Simon and Schuster, 1986.

27. Zukav, Gary. The Dancing Wu Li Masters. New York, New York: Bantam, 1980.

FILMS

28. The Ark.
29. The Emerald Forest.
30. E.T., The Extraterrestrial.
31. Goddess of the Earth, Public Broadcasting System.
32. The Gods Must Be Crazy.
33. Never Cry Wolf.

ABOUT THE AUTHOR

Since 1959, Mike Cohen, Ed.D. has organized consensus-based camping trips that explore in depth the ecology of North America. He is an environmental psychologist, a founder and director of the National Audubon Society Expedition Institute and conceiver of the 1985 Symposium IS THE EARTH A LIVING ORGANISM? He is also a member of the Adjunct Faculty, Lesley College Graduate School and of the Guild of Tutors, International College.

Dr. Cohen sings and plays traditional music, calls contra dances and enjoys photography, backpacking and cross-country skiing. He is a frequent conference speaker and his previous books include **Prejudice Against Nature: A Guidebook for the Liberation of Self and Planet** (1983), **Our Classroom is Wild America** (1974) and **To Hell With Skiing** (1966).

A READING <u>MUST</u> FOR ALL CARING PEOPLE

DIET FOR A NEW AMERICA
How Your Food Choices Affect Your Health, Happiness and the Future of Life on Earth
John Robbins

DIET FOR A NEW AMERICA is an extraordinary look at our dependence on animals for food, the profoundly inhumane and unhealthy conditions under which they are currently raised, and the physical, emotional and economic price we unknowingly pay for our eating habits. It reveals exactly what many of those responsible for producing the meat, eggs and dairy products we eat *don't* want us to know.

This beautifully written and impeccably researched book removes the smokescreen that has kept us ignorant about the food we eat. We are a society that prides itself in mass production; yet our indiscriminate overuse of pesticides and chemicals to feed the animals we raise for food exacts a truly incredible hidden toll on the health of every single one of us, as well as on the entire ecological system of our planet.

This powerful book not only addresses these staggering implications, but it shows the way out and provides suggestions for a new code of behavior that can greatly enhance our health, awareness and humanitarianism.

Paper, $10.95, 424 pages
photos, graphs, charts and bibliography
Hard Cover, $15.95

Endorsed by Harvey and Marilyn Diamond, authors of the bestselling book FIT FOR LIFE (over 2 million hard cover copies sold)

PULITZER PRIZE NOMINEE!